T0265690

Starved for Light

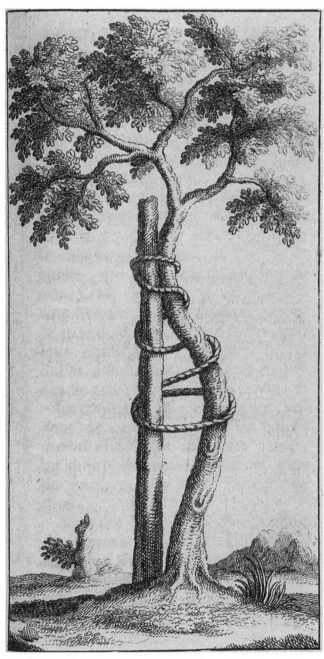

Frontis "The Andry Tree," an etching published in Nicholas Andry's 1741 publication *L'orthopédia*.

Starved for Light

The Long Shadow of Rickets and Vitamin D Deficiency

Christian Warren

The University of Chicago Press
Chicago and London

The University of Chicago Press, Chicago 60637
The University of Chicago Press, Ltd., London
© 2024 by The University of Chicago
Published 2024
Printed in the United States of America

33 32 31 30 29 28 27 26 25 24   1 2 3 4 5

ISBN-13: 978-0-226-15193-9   (cloth)
ISBN-13: 978-0-226-15209-7   (e-book)
DOI: https://doi.org/10.7208/chicago/9780226152097.001.0001

Library of Congress Cataloging-in-Publication Data

Names: Warren, Christian, author.
Title: Starved for light : the long shadow of rickets and vitamin D deficiency /
    Christian Warren.
Description: Chicago ; London : The University of Chicago Press, 2024. |
    Includes bibliographical references and index.
Identifiers: LCCN 2024013155 | ISBN 9780226151939 (cloth) |
    ISBN 9780226152097 (e-book)
Subjects: LCSH: Rickets—History. | Vitamin D in human nutrition—
    United States—History. | Rickets—United States—Prevention—History.
Classification: LCC RJ396.W37 2024 | DDC 616.3/95—dc23/eng/20240501
LC record available at https://lccn.loc.gov/2024013155

♾ This paper meets the requirements of ANSI/NISO Z39.48-1992
(Permanence of Paper).

Contents

Introduction

Sun and Skin and Bones

Just a dozen miles west of the George Washington Bridge, in Wallington, New Jersey, is an empty field where, for over forty years, the Farmland Dairy processing plant once stood. In its last years, after a major upgrade in 2006, the plant could process up to a quarter of a million gallons of milk each day, milk trucked into the thirty-nine-acre facility from dairies throughout the northeast. Between the pumps that drained the tanker trucks arriving at the facility and the pumps that carefully filled cartons and bottles with the finished product, the grade A liquid passed through a state-of-the-art facility larger than two football fields. In all such plants, technicians in lab coats, masks, and hair nets monitor the machines that test, filter, and pool the milk; skim the cream; and pasteurize, chill, reblend, and homogenize the resulting array of liquid dairy products. And into every batch, an infuser adds a carefully calibrated dose of "cholecalciferol," commonly known as vitamin D, "the sunshine vitamin."[1]

If the precision, the dedication to hygiene, sterility, and precisely calibrated quality control seem more appropriate to a laboratory or hospital than to a factory, let alone a farm, that is no mistake. There is more going on here than transforming raw milk into groceries; the plant is turning a dairy product into a drug-delivery device. Unlike pasteurization, homogenization, and other processes that remove impurities, kill microbes, improve shelf life, or simply make for a more palatable product, dosing milk with vitamin D does not restore some natural goodness lost in modern dairy processes. It adds no color, flavor, or odor to the milk. It is instead a public health intervention: each eight-ounce serving of "vitamin D milk" provides three or four micrograms of the nutrient, 25 percent of the amount recommended to maintain bone health in children and adults. The amount of vitamin D in raw milk at its source—the cow—varies wildly and in no case approaches the Recommended Daily Allowance.[2] So if cow's milk is not a natural source of this nutrient, why is universal fortification of milk America's preferred method for its delivery? More important, what precisely is the health problem that warrants adding a pharmaceutical into our milk supply every day?

The answer today is imprecise, tautological in fact: we fortify milk with vitamin D to prevent vitamin D deficiency—not so much a disease as a risk factor, implicated in dozens of conditions from osteoporosis to heart disease to ADHD. In the 1930s, by contrast, when American dairies began fortifying milk with the vitamin, the reason to do so was as precise as it gets: to prevent and cure rickets, the softening and weakening of children's bones that often produces bowed legs and other deformities.

The word "rickets" doesn't convey what it once did. In the popular imagination today, the word evokes a bygone era. It gets lumped in with other dreadfully misnamed "Victorian diseases"— antique, quaint, ghastly, and presumably cured.[3] Its power is also diminished by its place in the adjective "rickety," deployed to describe wobbly or decrepit chairs, cars, or houses: diminished objects, probably irredeemable. As for rickety people, the disease's most visible symptoms are bowed legs and knock knees, inviting our compassion, to be sure, but for centuries the root of countless jokes and songs, as recent as "Everyone Knows Juanita," who "stands in a bowlegged stance," from the 2017 Disney animated feature *Coco*.[4]

But to doctors a century ago, rickets was no laughing matter. In a 1926 issue of *The Delineator*, eminent pediatrician John

Howland described rickets as "one of the commonest of all diseases from which children suffer." He informed the women's magazine's readers that 90 percent of "city-bred" infants had rickets in the winter months, and that "the exaggerated bow-legs and the X-shaped knock-knees of the waddling children are familiar sights." But the effects went far beyond distorting children's legs. "It is hard to measure exactly the influence of rickets on mortality. It is surely considerable." Because the federal government did not collect such statistics, Howland could not have known that state agencies that year recorded 580 children felled by rickets. He was, nonetheless, confident about how rickets brought death: "The deformity which most seriously affects the general health is that of the chest." The rickety chest "sinks in," interfering with its normal expansion. As a result, the frequent infections of early childhood, especially those of the lungs, become "more than the child can bear." As to what caused rickets, he did not mention vitamins at all, though vitamin D had been discovered four years earlier. Perhaps assuming his lay readers would not yet be familiar with the new discoveries, he limited his discussion to sunlight and nutrition, two factors clearly implicated in rickets, as suggested in his article's title, "Starving for Sunshine."[5]

"Sunshine vitamin" is a popular nickname for vitamin D, and an appropriate one, even though, strictly speaking, it is not a vitamin at all, but a steroid hormone.[6] Chemists a century ago started isolating and naming the compounds that we now know by chemical names like 1α,25-dihydroxyvitamin D3 (1α,25(OH)2D3 for short). The discovery of dietary "magic bullets" from A to E outpaced organic chemistry's ability to define them (though not the market's ability to sell them), and "vitamin" conveyed just the combination of science and succinctness the new advertising age demanded.[7]

On the other hand, the "sunshine" in "sunshine vitamin" is perfectly apt. All that is required for our bodies to produce vitamin D is a few moments of exposure to the ultraviolet rays of the sun, which transform a basic sterol present in the skin of all higher animals into vitamin D3.[8] Though not the photosynthesis we learned about in biology class, the production of vitamin D in the skin *is* a form of photosynthesis—one we share with our evolutionary cousins in the plant kingdom. Millions of years before they grew a backbone, our ancestors relied upon vitamin D, though what part these steroid hormones played in early Paleozoic plant and animal metabolism is somewhat cloudy. With the rise of vertebrates, however, the vitamin took on a much clearer, and crucial, role: facilitating and regulating the absorption of calcium and phosphorus.[9]

The vitamin D manufactured in the skin undergoes two more transformations in the liver and kidneys. The resulting compounds act at the cellular and genetic level, affecting cell growth from blood to reproductive cells to bones and hair follicles. Their central role in regulating calcium and phosphorus levels in the blood makes adequate vitamin D imperative for bone growth and integrity.[10] Still, vitamin D can only work with the stores of calcium and phosphorus present in the body; it does not replace them. Rickets can occur if any of the three are deficient: a surplus of vitamin D will not compensate for a lack of calcium or phosphorus in the diet, nor will consuming extra calcium do away with the need for adequate vitamin D. Other metabolic disorders affecting phosphorus or acid levels in the blood can also produce rickets-like symptoms.[11] But by far, the most reliable path for otherwise healthy infants to contract rickets is to be shielded from the UV light of day and deprived of any dietary source of vitamin D.

: :

Both diet and darkness played their parts in the case histories of two baby boys, separated by a continent and by a century, but suffering from the same disorder. One, a six-month-old in San Francisco, "was believed to be in perfect health." His doctor, Lewis Mace, observed "his cheeks were fat and rounded, his expression intelligent and active, and his whole body plump and well proportioned." The other boy, a seventeen-month-old patient of Atlanta pediatrician Norman Carvalho, had a good appetite—his "daily caloric intake was 114 percent of the recommended dietary allowance." The first boy's concerned mother had brought him to the doctor for relief from a bout of constipation, the second's because he had stopped growing soon after being weaned. Both infants were found to have pronounced rickets. Careful physical examination of the first boy revealed the characteristic skeletal signs: an octagonal shape to his skull "with prominent frontal eminences," and a well-marked series of bead-like knobs on his ribcage. The second boy too exhibited this "rachitic rosary" among other skeletal deformities that left him unable to sit up or crawl.[12]

The decades between 1904, when Mace published his report, and 2001, when Norman F. Carvalho and his team of Atlanta pediatricians published theirs, brought impressive advances in understanding, preventing, and treating rickets. From one perspective, this legacy of advances maps splendidly onto popular ideas about medical history. It is a tale of trial and triumph, one disease after another falling before a phalanx of perspicacious epidemiologists and dogged scientists and doctors, all working within complex systems that enmeshed industry, academic medicine, government, and commerce. And indeed, the more recent report bears the markers of modern medicine, crisply ticking off the child's clinical signs, x-ray findings, and precise results of laboratory assays of mineral and hormonal levels in the child's blood; calmly describing treatment with injections of the long-established elixir, "ergocalciferol" (vitamin D2); and humbly summarizing the boy's rapid recovery as measured against growth charts and established "gross motor milestones."

The history of rickets is deeply intertwined with the emergence of modern medicine. This book will demonstrate that, time and again, rickets was a player in important chapters of that drama. Some of these roles will seem natural: the bowed legs of rickets survivors provided clinical material for the first generations of pediatric orthopedists; the search for the causes of rickets enmeshed it in the development of nutrition science, as well as the long history of health and the built environment. Other parts rickets played were less obvious if just as consequential. For example, the skeletal aftereffects of childhood rickets were a major factor in the rise of modern obstetrics, from the gradual takeover of midwifery by male physicians in the eighteenth century to America's reliance on cesarean delivery in the twentieth. Rickets is also entangled with the history of medical ethics. Thomas Percival, a founding theorist of modern medical ethics, reported ethical studies on rickety children in his Manchester infirmary clinic in the late 1700s; 130 years later American physicians ignored the basic principles of medical ethics in experiments to test cures for rickets. If rickets shared in the glory of medicine's fabled rise to a "golden age" of miracle cures, so too was it embroiled in the controversies and uncertainties that trouble today's wary and weary medical culture.

Moments of surprise such as those expressed by Drs. Mace and Carvalho, surprise often born of professional overconfidence, constitute one of this book's recurring themes. Both Mace and Carvalho frame their reports around a narrative of *surprise*: upon close examination of the seemingly well-cared-for middle-class child, the clinician diagnoses a disease historically associated with poverty. "These positive evidences of rickets were observed with much surprise in a child who had every advantage of attention, the best hygienic surroundings and the most scrupulous care," Mace remarked, and Carvalho reported that "the parents were well-educated, seemed knowledgeable and responsible, and had at least average family incomes."

Carvalho might also have experienced a second surprise— that *any* American in 2000 would have such pronounced rickets. Here we are, the fattest nation in the world, facing a resurgence of a disease of malnutrition. Here we have parents aggressively seeking health for their children by taking them "off the grid" of America's commercial food industry (where universal vitamin D supplementation is the norm), some of them slathering their kids with sunscreen or keeping them out of the sun entirely to avoid unnecessary cancer risk. And, ironically, applying the precautionary principle to cancer puts their children at risk of developing a disease long stigmatized through its association with poverty, ignorance, bad hygiene—and race.

Carvalho's report joined a growing body of late twentieth- and early twenty-first-century studies finding clear evidence that the number of rickets cases was climbing. It would seem that the ancient crippler, widely assumed to have been all but eradicated in the United States, was reemerging.[13] In the last decades of the twentieth century, researchers in medical centers from Alaska to Georgia, Philadelphia to Seattle, published dozens of studies on the prevalence of vitamin D deficiency and rickets. The case definitions of vitamin D "sufficiency," "insufficiency," and "deficiency" were inconsistent. Nevertheless, as one literature review put it, "findings from the recent literature in aggregate strongly suggest that hypovitaminosis D affects healthy children of all ages and all races."[14] Between one-fourth and one-half of children tested in many studies had "insufficient" levels of vitamin D, and significant numbers were "deficient."

Far less ambiguous were the hundreds of children with clinical rickets found in these studies, infants exhibiting some or all of the symptoms Lewis Mace noted in his patient a century before. As with the infant in Carvalho's care, almost all of the recent cases of rickets involved African Americans who were breastfed but who did not receive vitamin supplements; many of their parents had also severely limited their infants' exposure to sunlight. Many of the recent reports shared Mace and Carvalho's surprise to be faced with a case of rickets. That in itself is not surprising, given the broadly accepted "truth" that universal vitamin D supplementation of dairy and grains had vanquished rickets decades earlier.

A century ago, Mace and his generation in public health and the medical professions confronted an epidemic of rickets in American cities. Thirty years later, the general celebratory tone of the popular and medical press gave the impression that we were well on the way to the "cure." But was this impression even close to correct? One question for the historian is whether the end of the twentieth century brought an actual reemergence of rickets, or had this disease of diet and environment persisted through decades of complacence about a job well done, contradictory to the enduring but misguided faith that we had driven rickets back into the shadows through widespread vitamin D fortification?

The main outlines of an answer become obvious with even a preliminary survey of the medical literature: rickets was never eradicated. But even if that *were* true, the story of rickets and vitamin D deficiency is not just about bent bones. The relatively small number of cases of classic rickets today is merely the visible manifestation of widespread vitamin D deficiency throughout much of the world. Every month sees new clinical and epidemiological studies demonstrating low average vitamin D levels in the general public and linking those low levels to serious health risks, including cardiovascular disease, autoimmune diseases, and many forms of cancer. The number of prescriptions for blood assays of vitamin D is skyrocketing, as is demand for vitamin D supplements; more and more products at the grocery store proclaim that they are fortified with the "sunshine vitamin."

The more interesting questions, then, *assume* that rickets persists among a small portion of the population throughout the twentieth century, the visible sign of a much more widespread problem of vitamin D deficiency. In that case, what is it that drove rickets and chronic vitamin D deficiency from the shadows to prominence again? This book must examine how rickets was—and then was not—an important subject for America's doctors, public health institutions, and parents. But it does more, asking what is the relationship between the apparent resurgence of rickets and the growing concerns with measuring and increasing vitamin D levels in the general population? Vitamin D has taken center stage in a growing health debate over modern times that have us spending too little time out of doors; but now as throughout the last century, the most common solution is a biomedical fix: take your sunshine in a manufactured supplement!

The rise of today's pride and faith in supplementation, so widely held (though so thoroughly undeserved), makes for a fairly straightforward narrative arc from discovery to implementation to celebration. The subsequent loss of certainty about the risks and benefits of vitamin D comprises a more distinctly social, medical, and political historical narrative and is the heart of this book. After finishing it, you will know why no one should be surprised that vitamin D deficiency and rickets are "back."

Rickets has often been seen as more a nutritional than an environmental disorder; no surprise then that both Drs. Mace and Carvalho found food (the wrong kind of food, not the lack of it) to be the root cause of their patients' condition. The *real* problem was, they concluded, special food that their health-conscious parents fed them, but which lacked important health-giving nutrients. Mace's patient had been fed "a proprietary infant food" almost from birth; this formula, "warranted to produce a fat baby in every instance, had done so in this case, but the very presence of the fat had concealed the real condition of affairs." Carvalho's patient had been breastfed until nine months, after which he was "weaned to a soy health food beverage, which was not fortified with vitamin D or calcium." Neither report questioned the parents' devotion to their children; instead, they criticized their misplaced faith in foods that were sold as healthful and natural but which were neither. Both reports blame a lack of cow's milk in the boys' diets, but for different reasons: Mace assumed that "natural, raw cow's milk" would have provided the protein and fats whose antirachitic benefits were "too well known to require comment"; Carvalho on the other hand, knew that milk itself is not a rickets preventive, but that the vitamin D added to nearly all milk sold in the United States *is*.[15]

The physicians in both case reports make no apologies for their criticism of the dietary choices made by their patients' parents. Within any community, those who depart from dominant foodways or childrearing norms risk the harsh judgment of their peers, no matter that they defend their "deviance" as enlightened, progressive, or more healthful than the status quo. Even more tension arises when outsiders—whether from the local church, the health department, or Washington, DC—seek to impose their "improvements" in either childrearing or dietary practices upon the communities they purpose to help.

If, from the perspective of the medical establishment, rickets was in large part a culturally bound dietary disorder (as opposed to an environmental one), then its cure might best come through a scientifically improved diet. Supplementing milk and grains with industrially manufactured cholecalciferol became the American cure, just as iodine, thiamin, niacin, riboflavin, folic acid, lysine, and fluoride came to be added to the nation's food and water to prevent other specific ailments.[16] American health experts' decision to use dairy products as the delivery vehicle for mass vitamin D supplementation was a cultural decision, not a strictly scientific one, demonstrating the overlay of culture and nutrition science. Time and time again in the history of rickets, outcomes that seemed inevitable turn out to be the product of culture-bound decisions—contingencies, not fate.

In their focus on diet (with its implicit criticism of the parents), Drs. Mace and Carvahlo downplay the unhealthful physical environments in which their rickety patients lived. For most of human history, the vitamin D to build strong and straight bones came less from one's diet than from everyday exposure to the sun: working, doing the marketing, socializing, or playing took place in much greater measure in the open air; even the miner who toiled six long days in sunless depths spent much of his Sabbath out of doors. Two fundamental changes in the built environment over the last century and a half drastically altered this dynamic. First, the scale and density of our cities worked to blot out those rays of light not already trapped in the coal smoke of home and factory. Second, our homes and buildings became more commodious and self-contained—both functionally, allowing more and more activities to be undertaken indoors, and self-contained in the sense that buildings became increasingly shut off from the world outside, through ventilation, artificial lighting, and, eventually, air conditioning.[17]

Sages a century ago, such as Theobald Palm, foresaw that this rapidly evolving environment was itself the danger: "The exhilaration which we experience in bright weather," he wrote, "and the depression caused by a continuance of dull gray skies, must have a corresponding physical change in the nutrition . . . of our nervous centres, and . . . upon general nutrition." The trend, a British advocate for fresh air and exposure to nature warned, was to live "ever more and more in darkness and asphyxia."[18] By midcentury, reversing that trend appeared impossible. In this rapidly modernizing era, any hope of preserving the ancient relationship between sun, skin, and bone was discarded as quixotic if not retrograde. In the move into the modern built environment, we traded the unpredictability and bother of sunlight and natural ventilation for the control and reliability that modern electric lighting and ventilation systems promised. This meant, however, that new means of providing the "sunshine vitamin" were needed.

: :

To explain how dietary vitamin D came to replace sunlight, this book simultaneously explores two very different histories. One takes the long view, tracing cultural, environmental, and even evolutionary forces, weaving together migration, geography, ethnicity, and other stories of particularly long duration. These histories enliven and inform this book, but they are not a primary focus after the first chapters. Instead, these tales of ancient origin shape the book's central story: the dramatic and comparatively sudden change in how modern Americans secured the vitamin D they needed to build straight, strong bones and healthy bodies. *This* story has a reasonably clear narrative, with a beginning, middle, and, if not an *end*, then certainly a unique present. This history is driven forward by identifiable crises, scientific discoveries, and academic and industrial rivalries; what is more, it ends with a most satisfactory ironic outcome, as even the elixir developed to supplement (or indeed replace) the health-giving sun came to be replaced by industrially produced, commercial, and medicalized chemical products. In whatever form, vitamin D products seemed to hold the promise of complete dietary control and the mitigation of at least some of the effects of unhealthy environments. What this story lacks, in terms of framing a ripping narrative, is the absence of a central villain, though there is no shortage of bad actors, institutional blindness, corporate manipulation, and a steady drone of cultural and institutional racism.

The first two chapters trace the history of medical understanding of rickets, from the ancient world to the early twentieth century, when the modern understanding of vitamin D began to emerge. The first published European accounts in the seventeenth century found rickets among pampered middle- and upper-class children. The rise of the Industrial Revolution reversed this impression of rickets as a malady of wealth; under Britain's darkening skies, rickets transformed into the "English Disease," contorting the bodies and shortening the lives of children raised in the grinding poverty and sunless cityscapes of Blake's "dark satanic mills." Meanwhile, in the sun-baked American South, the inhuman conditions under the slave economy inflicted similar burdens on African and African American children, feeding the trend in the United States for race more than class to dominate conversations about rickets' victims. Toward the end of the nineteenth century, America's booming urbanization and industrialization produced environmental and social conditions like those of Dickensian England. American physicians and public health advocates turned their attention to rickets among urban immigrants, both African Americans and southern Europeans.

Through the early twentieth century, rickets pointed out vast gaps in knowledge about basic biochemistry and physiology. Chapters 3 and 4 explore how scientists came to understand the underlying mechanisms of vitamin D, and how they resolved the question of whether something in the diet or something in the environment was to blame for rickets. A parallel line of experimental inquiry focused on therapeutics and public health—the social causes of rickets, its natural history, and preventive measures. Some of these studies involved experimentation on institutionalized infants, a troubling legacy that clouds the substantial progress of the day. This history falls smartly in line with the critical historiography of the "Golden Age of Medicine" and its counterpart, the

"New Public Health."[19] In the same years, thousands of children untouched by preventive measures, their bones curved and distorted by rickets, underwent painful surgery, or were fitted with orthopedic devices to correct or compensate for the bodily harm done by inadequate diet and want of daylight.

Chapters 5 and 6 trace the history of dietary sources of vitamin D, from cod liver oil to today's potent capsules of the sunshine vitamin. Atlantic fisherfolk have been swallowing cod liver for centuries, and long before anyone had a notion of vitamins, cod liver oil was a popular health tonic for young and old. Preventing rickets was just one of its many uses, and parents forced their youngsters to swallow the nasty-tasting oil long before the 1920s, when scientists identified vitamin D as its active ingredient. Then, with crucial support from the food industry, those scientists figured out how to produce the chemical in the laboratory (and subsequently the factory), creating the basis for the modern solution to rickets—"universal" supplementation of dairy products with vitamin D.

From the 1930s, despite clear evidence of rickets' persistence (albeit at lower and lower rates) and ongoing research into its prevention and cure, there rose the myth that rickets in the United States could be consigned to the history books. Chapter 7 explores the nature of this partial blindness and its heavy costs. Chapter 8 revisits the apparent resurgence of rickets toward the end of the twentieth century in both the United States and in the rest of the world. It then turns to the ancient roots of vitamin D deficiency, tracing the role of sun, skin, and bone in shaping human evolution, and discussing the controversial place rickets retains as a "race disease." In recent years vitamin D deficiency evolved from mere precursor to a disorder—rickets—to its current definition as a disorder itself, implicated in increased risk of all manner of conditions from depression to cancer. In the modern clinical setting, uncertainties about the value of diagnosing along racial categories (or uncertainty about whether such categories even exist) blend with uncertainties about "low" levels of vitamin D, however "low" is to be defined.

∷

The history of rickets and vitamin D deficiency—aspects of it, in any case—has attracted scholars of many generations, although that abiding interest has not translated into major studies. Broad surveys have appeared as chapters in books on nutrition, orthopedics, and medical history encyclopedias. The lineage of the Greats—from Whistler and Glisson, to Mellanby and Steenbock, to Eliot and Hess—has been told in more or less hagiographic tones in countless medical and scientific journals.[20] Social historians of medicine have found much in the history of vitamin D and rickets instructive: Rima Apple examined the history of vitamin D as part of her studies of the science and business of supplements; Susan Lederer explored the ethical issues around early twentieth-century studies of rickets in institutionalized children.[21]

Several historians have put rickets itself—the dynamics of vitamin D, sun, skin, and bone—at the center of studies exploring medical research, industry, and the state. Roberta Bivins has been studying the British experience in the changing face of rickets over the twentieth century. M. Allison Arwady's Yale thesis on early twentieth-century clinical studies by Martha Mae Eliot and others provides a detailed assessment. Daniel Freund's *American Sunshine* analyzes how vitamin D and rickets impelled much of the early twentieth century's entrepreneurial and engineering dynamism aimed at enhancing (or replacing) exposure to the sun.[22] This book also draws upon a growing body of studies on the history of child nutrition—infant feeding especially, and the role of cow's milk as a "perfect food."[23]

Because rickets has fallen so heavily upon African Americans, even when the vast majority lived and toiled under the southern sun, the topic of vitamin D deficiency and rickets frequently appears in histories of health and nutrition of Black Americans.[24] Some have focused on nutrition, while others, examining the interplay of race and gender in the history of obstetrics in the nineteenth century, have explored the deadly stamp childhood rickets left upon the bones of African American women.[25] Economic historians have taken rickets' measure in their assessments of the slave economy, while medical anthropologists have directly measured the physical toll of rickets in bones unearthed from cemeteries around the country, verifying the conclusions historians reached from the written record.[26] A number of historical studies have investigated nutrition and infant feeding choices of African American families from the era of emancipation. African American women's reembrace of breastfeeding in recent decades has been celebrated by nutritionists and pediatricians, but it also played an important role in the late twentieth-century resurgence of vitamin D deficiency and rickets. Also important has been a decline in the popularity of America's chief delivery device for vitamin D fortification, cow's milk.

::

Milk consumption had been falling steadily long before 2013, when the massive Farmland Dairy complex in Wallington, New Jersey, shut down its pumps and pasteurizers.[27] Farmland was not the victim of cutthroat competition from behemoths—it was itself the property of one such multinational agribusiness—as much as just another casualty of Americans' declining demand for milk. In the late 1960s, per capita consumption stood at thirty gallons per year. It plummeted to under sixteen gallons in 2021.[28] The reasons for the decline are as numerous as the variations on the ubiquitous "Got Milk?" slogan of the 1990s, now itself relinquished to the dustbin of advertising history: from diet consciousness, the rise in reported rates of lactose intolerance, veganism, and humanitarian and environmental concerns, to social and political opposition to multinational food industries. In the natural scheme of things, this turn from milk would have nothing to do with vitamin D deficiency and rickets and have no place in this book. After all, natural milk is a poor source of vitamin D. But because almost a century ago milk came to be seen as the "natural" vehicle for delivering vitamin D in the United States, lower milk consumption means one more cause for ever-lower baseline vitamin D levels. Will we find a different basic foodstuff to fortify, or move to a different model for delivering the "sunshine vitamin"? It hardly seems likely that we will turn back to bribing sour-faced children to swallow their daily cod liver oil. Perhaps we can reshape the built environment to produce again the conditions that prevailed before vitamin D deficiency started softening our infants' bones. But how far back in history must we turn for our model? When and where was that Golden Age of sunshine and straight legs, if it ever existed?

"Coeval with Civilization"

Rickets from Ancient Egypt to the "Dark Satanic Mills"

The Francis A. Countway Library rises five stories above Hunting-ton Avenue, four lanes brimming with Boston drivers, bisected by the steel rails of the Green Line, the oldest of Boston's streetcar lines. Built on the southernmost corner of the Harvard Medical School campus, the Countway is easily accessible to the thousands of health professionals and students of the hospitals, laboratories, education centers, and clinics of the Longwood Medical Campus.[1] After visitors pass through a spacious atrium, they can take a quick elevator ride to the sunlit fifth floor, once the home to the Warren Anatomical Museum. More than a few times when I was doing re-search at the Countway years ago, I rode that elevator and mean-dered with students and tourists as they took in the collection's most visually compelling artifacts, painstakingly displayed in xenon-lit glass cases.[2] These curious specimens performed the old accustomed functions of every anatomical collection: to inform, surely; but also to scold, shock, or inspire contemplation—on the progress of medicine more than on every man's fate (presumably, doctors have little need of memento mori).

Second in popularity only to the gruesome skull of Phineas Gage and the forty-three-inch tamping iron that perforated his head was the "rickets skeleton" as described here in the collec-tion's 1870 catalog: "1544. Articulated skeleton of an adult female, showing the effects of rickets. From Europe. Pelvis and long bones of the lower extremities very greatly deformed; the position of the sacrum being horizontal; and the broad, thin, sickle-shaped form of the fibulae very characteristic. Strong lateral curvature of the spine. Head, upper extremities, and feet well formed; and the thorax very nearly so. Structure of the bones perfectly healthy. 1857. Museum Fund."[3] The skeleton presents a most fearful asymmetry, its twisting radii, femurs, and spine evoking the steel girders and beams of a Frank Gehry building under construction, defying symmetry, right angles, parallelism, and, it would seem, gravity. The viewer strains to imagine the pain and debility afflicting the soul whose body these misshapen bones carried. But this impulse to form an empathic bond is undercut by the overwhelming meta-narrative of the museum setting. With its careful display of relics, crisply worded object labels beside each, their language framed almost always in the past triumphant, the museum and the medical library housing it are a shrine to—and engine of—the very medical progress that the visitor assumes has relegated the pain of rickets to the history books.

And indeed, extreme cases such as #1544's are unheard of today and presumably were rare in 1857, when surgeon John Collins Warren Warren collected this specimen. If not, why would he pay the American physician who brought it back from Paris $89.10 for it? Why else would a French pathologist have gone to the trouble to mount it so carefully?[4] And were the conditions that produced the young woman's twisting frame unique to Victorian-era Paris?

To the last question, both the words and the bones of our ancestors answer with a qualified no. Ample written and physical evidence demonstrates rickets' presence in every age, in smaller or larger measure. The first published European accounts in the seventeenth century focused on the children of the more prosperous classes, but industrialization, urban growth, and migration of the nineteenth century drastically altered the "traditional" picture of rickets. In the darkening cities of Europe's Industrial Revolution, rickets, known by then to many as the "English Disease," contorted the bodies and shortened the lives of the children of the poor,

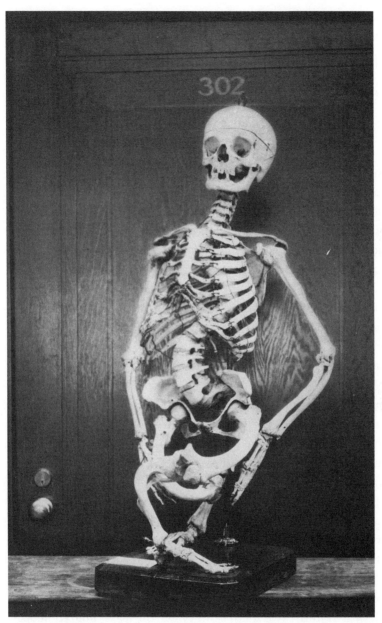

Figure 1.1 Skeleton of an adult female, showing the effects of rickets. Although rickets this pronounced has always been rare, the unmistakable aftereffects of childhood rickets were plainly visible in many people raised in the dark cities of the nineteenth century. Warren Anatomical Museum Collection, Center for the History of Medicine in the Countway Library of Medicine, Harvard University.

whether little English boys like Dickens's Tiny Tim or French girls
like #1544. The magnitude of this eruption of rickets recast it as
essentially a disease of poverty, though in reality what had arrived
was a new "normal" in which children rich and poor were deprived
of the sunlight needed to produce adequate vitamin D.

: :

In 1922, well into the modern era, British physician and health
administrator John Lawson Dick published *Rickets*, apparently in-
tending it to be the definitive book on the subject.[5] But Dick's views
on the causes of rickets were out of step with those of his generation, with its energetic and monomaniacal quest for the nutritional
equivalent of magic bullets—unique substances whose absence
from the diet triggered specific health problems. Dick dismissed this
as reductionist. "The vitamin theory of disease is a fascinating one,"
he sniffed, "and is one of those facile hypotheses which will explain
anything."[6] His abiding interest in social medicine proceeded from
long experience with military medicine, from service in the Boer
War to a post in London's military hospital during the Great War.
In addition, his postwar service in the Ministry of Pensions and as
president of the London medical boards convinced him of a clear
connection between environment and all manner of disease. The
curved limbs of rickety children were likely to appear, he believed,
wherever one found the "right" environmental factors: "Vitiated
atmosphere in close and confined dwellings; . . . the exclusion of
sunlight; . . . the lack of opportunity of exercise; . . . damp climates
and long winters."[7] For evidence of this, Dick surveyed the health
literature from around the world and the published history of rickets from prehistory to modernity. He wanted to demonstrate that
rickets had been rare in antiquity and that in the twentieth century CE as in the twentieth century BC, the least civilized peoples
enjoyed the greatest freedom from rickets.

Three generations of mania for exploring the ancient world,
exemplified by Heinrich Schliemann's quest to discover the historic Troy, Arthur Evans's search for Knossos, Giuseppe Fiorelli

uncovering Pompeii, and Howard Carter's explorations of the Valley of the Kings, produced a wealth of material for postmortems conducted by paleontologists and historically minded physicians. With the 1895 discovery of the x-ray, they had a particularly powerful new tool. Soon, paleopathologists from Berlin to London were carting ancient bones and mummies to the nearest radiograph machine—none more colorfully than famed Egyptologist Grafton Elliot Smith, who hauled a mummy he wanted to x-ray through the streets of Cairo in the back of a taxi.[8] Before long, large-scale projects of wholesale mummy radiology were underway.[9] And while rickets was not their prime concern (they were far more interested in tuberculosis and syphilis), any evidence of the distorted bones of rickets could hardly escape their notice.

Even so, little if any evidence of rickets appeared in the thousands of x-ray plates, or among the physical remains of the tens of thousands of mummies brought to autopsy.[10] It was clear to Dick that rickets "was unknown in the ancient civilisations of Babylonia, Egypt, and India," a finding at odds with the assertion, common in his time, that "rickets is a condition coeval with civilization."[11] But Dick easily sidestepped the apparent contradiction by pointing to the globe. "The earlier civilisations were cradled in kindly nurseries," their tropical latitude and sunny climate mitigating much of the harmful effect of urban density and crowding.[12] Far more powerful in explaining why the ancient civilizations failed to produce this archetypal "disease of civilization" was the vast gulf separating ancient and modern cities in the scale of their economies and the organization of labor. Rather than a consequence of civilization, Dick concluded, "it would be more accurate to say that, considered from a historical point of view, [rickets] is a disease of industrialism."[13]

We have stronger evidence of rickets from the ancient cities on the northern shores of the Mediterranean, with both the medical literature and archaeological evidence suggesting that vitamin D deficiency was a regular visitor in Athens and Rome. In the second century CE, Galen mentioned bony deformities suggestive of rickets. Soranus of Ephesus was more explicit. He observed that bent legs, curved spines, and bony protuberances were far more common in infants from the Roman capital than those raised in the country, but he blamed nurture far more than environment, asserting that "the women of this city do not possess sufficient devotion to look after everything as the purely Grecian women do." Without proper supervision of infants, "the limbs of the majority become distorted, as the whole weight of the body rests on the legs." Environment played its part in Soranus's analysis, but not in the way that John Dick's sun- and air-obsessed generation would have considered. The full weight of Roman infants was born by "solid and hard" stone pavement. And, Soranus argued, "whenever the ground upon which the child walks is rigid, the imposed weight heavy, and that which carries it tender—then of necessity the limbs give in a little, since the bones have not yet become strong." The limbs of rural children, he implies, did not have to absorb this abuse, as the child of the country walked upon soft grass and pliant soil. With our modern understanding of vitamin D, sunshine, and rickets, we might conclude that Soranus reached the correct conclusion—about the benefits of country life, not mothers' devotion—for the wrong reason.[14]

The ancient record does not support Dick's assertion that rickets arose as a consequence of "civilization" *or* industrialization. Instead, latitude and especially urban density seem more predictive, associations that persist across the Middle Ages and well into the Early Modern period. Archaeological studies of medieval cemeteries from English towns to the Baltic port city of Schleswig have found small but significant instances of infant remains bearing the unmistakable signs of severe vitamin D deficiency.[15] The well-preserved remains in these studies were usually recovered from coffins, more evidence that prior to the Industrial Revolution rickets disproportionately affected the children of the wealthy. The 2004 discovery of a hidden crypt beneath the floors of Florence's Basilica of San Lorenzo led to yet further support for this impression. The crypt held the skeletons of nine infants, ranging in age from newborn to five, born to the Renaissance dynasty, the Medicis. Six of the skeletons bore the unmistakable signs of rickets, with the oldest, Filippo, exhibiting not only bent long bones, but also a visibly distorted "rachitic skull."[16] Without corresponding studies from potter's fields and bone houses, the degree to which rickets was a disease of the comfortable remains unverifiable. One such set of bones emerged in 2013 when a London work crew drilling a twelve-foot-wide ventilation shaft for a new subway tunnel uncovered the graves of two dozen fourteenth-century Londoners. The facts of their lives as discernible from their bones and the nature of the burials indicate they represented London's poor. A significant number of these skeletons showed clear evidence of rickets.[17]

Not all of the rachitic children of the sixteenth and seventeenth centuries went unremarked in their time, even if their condition did not yet have a name. Strong consensus has long held that the visible deformities and halting gait of King Charles I of England were the result of rickets. His daughter Elizabeth exhibited even more severe stigmata, with a badly curved spine and bowed femurs that produced knock knees.[18] The princess died at the age of fifteen, only nineteen months after Charles was beheaded at the behest of the Rump Parliament's high court. The cause of Elizabeth's death was considerably more ambiguous than her father's. Her doctors reported that she died of pneumonia. She had been sickly from infancy, always prone to respiratory complaints—another signal complication of severe rickets.[19]

The rickety king and the sickly children born to a powerful family possibly felled by a disease of malnutrition, make for compelling stories. More significant to the history of rickets, though, are seventeen London children whose names (and bones) are lost to us, but whose deaths were noted in the city's 1634 Bill of Mortality under a new cause of death—"rickets," the word's earliest known appearance in print.[20] Historians of medicine ascribe great significance to the naming of diseases, even apart from the medical profession's tradition of naming conditions after the doctors who first identified them.[21] None of the cases discussed so far in this chapter, from Soranus's bowlegged Roman children to the young Medicis, were diagnosed with a specific disease.

To the degree that naming a disease creates it, rickets did not exist before the mid-seventeenth century. Most published accounts of the "discovery of rickets" give that honor to English physician Francis Glisson, whose 1650 *De Rachidite, sive Morbo Puerlili, qui Vulgo The Rickets Dictur Tractatis* was widely read, quickly translated into English (as *Rachitis, the Children's Disease Known in the Vernacular as the Rickets*), and reprinted three times in Glisson's lifetime.[22] Officially, bragging rights to first publication belong to another Englishman, Daniel Whistler, who, five years earlier, published his doctoral thesis, "*De morbo puerili Anglorum, quem patrio idiomate indigenae vocant the Rickets*" ("The English Children's Disease, Which in the Vernacular Is Called *The Rickets*"), at the University of Leyden.[23] But Whistler's slim thesis was largely ignored, and some historians have questioned its originality.[24]

Glisson's book was officially a group effort, bearing the names of several members of the College of Physicians with whom Glisson corresponded. But Glisson was clearly "first author." Trained at Cambridge, Glisson was made Regius Professor of Physic there in 1636, though most of his professional life was spent in London. With the outbreak of the Civil War in 1642, Glisson fled to the town of Colchester, where he spent much of the rest of the decade building a considerable practice and undertaking his study of rickets. Soon after the siege of Colchester in 1648, he moved back to London. In 1650 he published *De Rachidite*, the product of considerable clinical experience as well as impressive pathological study. Glisson brought dozens of children to autopsy, identifying far more than the skeletal effects of severe vitamin D deficiency. Because he suspected a dietary origin for rickets, he was particularly observant of pathologies of the internal organs. Whistler, too, suspected that diet was chiefly to blame for rickets.

Another thing Whistler and Glisson agreed on was the importance of naming a new disease properly, and that meant coining a solidly Latin or Greek name. Whistler went with *paedosplanchnosteocaces*, a pretentious inkhornism that never caught on. Glisson's far catchier *rachitis* became the term of art, enhancing his reputation for having discovered the condition.[25] How Glisson chose the name *rachitis* goes to the irony at the heart of these stories of "discovery" *and* of naming the disease. Glisson and his writing partners, knowing that the academic medical community "may peradventure expect a name from us," sought likely candidates in the glossaries of Greek and Latin. One of the team "fell upon a name" they could all agree on: the Greek word *rachitis*, meaning "disease of the back." The name made sense, given the frequent spinal effects of severe rickets; but perhaps more important, Glisson argued, "the Name is familiar and easy"—familiar because of its similarity to "the English name rickets," which seemed to be an old and customary word.[26] At this point, the authors make a remarkable linguistic leap, speculating that the English "rickets" may have derived from the Greek *rachitis*: "Without any wracking or convulsion of the Word, the name Rickets may be readily deduced from the Greek work Rachitis."[27] "Without any wracking . . . of the word" invites speculation that Glisson and his collaborators were playing on the multiple meanings, and perhaps etymology, of "wrack," alluding to both the device of torture and the vernacular "rick" (which derives from the Scandinavian *rykk*), a twisted or sprained limb.[28] Whether or not we can agree on the origins of the word "rickets," the fact that the titles of both treatises contained the vernacular "rickets"—pointing out, in fact, its common usage—suggests the word had long been used to name at least the skeletal aspects of severe vitamin D deficiency.[29]

If so, why did both Whistler and Glisson assert that rickets was, in Glisson's words "absolutely a new Disease, and never described by any ancient or modern writers in their practical books"?[30] Is the sudden emergence of rickets in the medical literature merely the consequence of a few alert and curious physicians, or was there a visible rise in rickets in seventeenth-century England? A number of factors in the seventeenth century may have conspired to produce a notable increase in rickets, especially visible among the children of the well-off who were Glisson's and Whistler's patients. John Lawson Dick argued that rickets "probably began to be prevalent in England and on the Continent some time toward the end of the sixteenth century," brought about by increased urban density and the smoke from coal fires.[31]

The longer and colder winters of the Little Ice Age, which reached its nadir in the late sixteenth century, undoubtedly played a part as well. Winter's grip lasted longer; days grew colder; in London the River Thames froze over with unprecedented frequency—nine times in the seventeenth century; grain output dropped precipitously, reducing average Europeans' stature, and catalyzing—it has been argued—the search for supernatural malefactors. While it may be plausible that cold weather fanned the flames of witch hunters, it certainly increased the need for real fires to keep warm.[32] Coping with the cold naturally increased smoke and soot and encouraged clothing and behaviors that would further reduce exposure to solar radiation. And while we might reasonably expect that prolonged cold weather affected the poor disproportionately, poor and wealthy alike experienced drastically reduced UV exposures.

None of this should have produced a higher incidence of rickets among the affluent, as Glisson and Whistler found, but the rickety children Glisson studied came from England's wealthier southern countryside; and their parents, Glisson noted, engaged in "a delicate kind of life abandoned to ease and voluptuousness, slothful, and rarely accustomed to danger, and care."[33] Whistler, too, perceived an explicit class overlay, finding rickets "most frequent in the ranks of the highest citizens, next amongst the dregs of the populace, least amongst those of moderate means."[34] Glisson advocated that children receive more light and air to prevent and cure rickets, and while it is tempting to credit him with foresight into the modern etiology of rickets, his recommendation drew more from the day's humoral system than from any specific observations he had made linking sunlight and bone growth.[35]

Cultural practices among elites played a crucial role in encouraging rickets among Glisson's well-heeled patients. In England, "the ranks of the highest citizens" included an increasing number of moneyed commoners, many eager to distinguish themselves from the ranks of working commoners from which they so recently rose.[36] They sought to effect the pale skin and blue veins of the aristocracy, whether cosmetically or by abjuring the sun that darkened the worker's complexion.[37] Many kept house in the city as well as homes in the country. In his early-eighteenth-century book on the history and benefits of cold-water bathing, Sir Baynard Floyer observed that rickets most affected "those that were rich and opulent, and put their children out to nurse." Floyer does not speculate whether this choice was made for the health of their infants or to enhance the parents' social opportunities.[38] Perhaps the increased smoke of the city added to the perceived advantages. He also praised Welsh women for routinely bathing their infants in cold water, since "by this the Welsh women used to prevent the rickets in their children; and 'tis a common saying among their nurses, that no child has the rickets unless he has a dirty slut for his nurse."[39] Life in the country would seem to guarantee these infants plenty of sunshine (hence vitamin D). This paradox of well-off city children developing rickets in the sunny countryside disappears with fuller consideration of the vitamin D-calcium metabolism and the probable health status of these nurses. Recent studies argue that the problem for these infants may not have been a deficiency of vitamin D but of calcium. Years of serial or multiple nursing could easily deplete a wet nurse's stores of calcium, putting her charges at great risk of rickets.[40]

So changing climate and the childrearing habits of the rising middle class provide plausible arguments in favor of Glisson's claim that rickets was an "absolutely new disease." The opposing argument, that rickets was not new, but only newly of interest, rests upon two categories of "testimony." First, there were the ample evidences visible to the physician and his countrymen, had they looked for them: linguistic evidence (some of which Glisson himself acknowledged) and folk healing practices, from herbal concoctions and food specifics such as raven's liver, to extended months of swaddling infants to prevent soft legs from bowing, to magic cures—passing "ruptured" infants over the cloven trunk of a sapling, for example. The second category is the evidence that we may safely assume should have been visible in the bodies of children, evidence that modern scholars have gleaned from burrowing through old public archives and private receipt books, inspecting the bones exhumed from tombs and boneyards, and generalizing from ample examples from other lands and times past and present.

Regardless of how common rickets was in England before Glisson's *De Rachidite* and whether he should be given credit for its "discovery," Glisson's work surely impelled the flood of published research on rickets in succeeding generations. Rickets was christened "Glisson's Disease," and on the continent it quickly came to be called the "English Disease."[41] Physicians in every generation after 1650 explored its causes, tested cures, measured its incidence, and dealt with the long-term consequences of bodies left misshapen. In 1666, the political philosopher and physician John Locke, then a medical student, "cut up a child of Dr. Aylworths dead of the Rhickets," finding serious heart deformities he assumed were a symptom or the cause of the child's rickets.[42] In the nineteenth century, rickets drew the attention of such medical luminaries as Rudolf Virchow, who compared childhood rickets with osteomalacia, its counterpart in adults, and John Snow, fabled breaker of the Broad Street Pump. There was even a medical dissertation by William Osler, not the famous Canadian, but a Scottish physician with that storied name.[43]

Between Glisson's time and the early twentieth century, when John Dick conducted his survey, treating the symptoms and sequelae of acute rickets played a foundational role in the rise of modern obstetrics and orthopedics, subjects to which we will return later. These centuries also saw rickets' transformation from a private tragedy of the "middling sort" to a major public health crisis centered in the poorest quarters of countless cities. Dick asserted that rickets of his day was a "disease of industrialism." If so, the transformation should have occurred in tandem with the rise of the Industrial Revolution in England's cities.[44]

: :

The poet and painter William Blake, for all the fantastical imagery and mysticism of his poetry and art, was as sharp a witness and critic of the Industrial Revolution as any of his contemporaries, recording in verse and image the deep cultural forces forging the new world and its effects, whether the coal spreading over "Every black'ning Church" or the plight of child laborers, such as the chimney sweep: "A little black thing among the snow: / Crying weep, weep in notes of woe!"[45] He saw the cities' residents crammed into cells, "every house a den, every man bound," and he saw the bodily burdens of this environment: "The citizens, in leaden gyves the inhabitants of suburbs / Walk heavy; soft and bent are the bones of villagers."[46] And in the preface to his epic poem *Milton*, he coined one of the most stinging terms for his age.[47]

The rise of Blake's "Dark Satanic Mills" produced conditions in England and other parts of Europe that John Dick's generation would later blame for the near endemic rates of rickets so visible in their time. The environmental changes over the two hundred years after Glisson were the product of more than simple growth. Ample commentary about smoky British cities extends back to the seventeenth century and earlier. London was particularly crowded—and packed into a walled city, affording its children far less sunshine than ancient Rome, where Soranus apparently found rickets a common condition. But as bad as British cities might have been in Glisson's time, by the mid-nineteenth century, the urban population had soared as bedraggled migrants from the countryside crowded into the slums of the sunless capitals of industrialization and proletarianization.[48] The smoke that enshrouded these cities, the filth and pestilential vapors of their neighborhoods, and their grinding poverty and deadly workplaces robbed these Britons of health, shortened their lives, and left the indelible stamp of debility and deformity upon their bodies. These deteriorating cityscapes drew frequent comment from social critics, while physicians, alerted to the signs and consequences of rickets, found its incidence on the rise. A few observers correctly associated the two trends, noting that the incidence of rickets was rising in lockstep with the smoke and squalor. For once, rickets among the poor did not escape the notice of the elite observers, who were set on documenting the true costs of their hard times.

Manchester was already an important center of cotton thread production in the mid-eighteenth century, when it began a half century of unguided growth as the old domestic "putting-out" system of production gave way to coal-burning, steam-powered mills.[49] The results impressed Alexis de Tocqueville when he visited there in 1835. "A sort of black smoke covers the city. The sun seen through it is a disc without rays." De Tocqueville linked the city's degraded environment with its degrading working conditions: "Under this half daylight 300,000 human beings are ceaselessly at work."[50] Friedrich Engels arrived there in 1842; for him the city epitomized "the degradation to which the application of steam-power, machinery and the division of labour reduce the working-man." These grinding forces left their mark on the city's children in the form of rickets. "I have seldom traversed Manchester," Engels recalled, "without meeting three or four [children] suffering from . . . distortions of the spinal columns and legs." The signs were "evident, at a glance. . . . The knees are bent inward and backwards, the ankles deformed and thick, and the spinal column often bent forwards or to one side."[51]

Some commentators touched on the effects of rickets as a familiar marker of debilitating English urban environments or working conditions, as writer and activist Flora Tristan did in "Promenades in London." Britain's factory workers were "shut up twelve to fourteen hours a day in mean rooms," breathing all manner of hazardous substances, subsisting on a diet that frequently passed "from insufficient nourishment to excessive drinking." The end result: "unfortunates" who were "pale, rickety, and sickly," with thin, feeble bodies with weak arms, wan complexions, and dull eyes."[52] The appearance of "rickety" in Tristan's description warrants comment. Was this the English translator's choice for Tristan's word meaning "unsteady, or wobbly"? Consulting the original French edition confirms that Tristan did indeed describe the workers as "*étiolés, rachitiques, souffreteux.*"[53] It is still not clear whether she was presuming to make a clinical diagnosis or merely using *rachitiques* to evoke the workers' debilities; similar ambiguity arises in a passage elsewhere, where she says of children in London's slums, "What could be more rickety, more cadaverous than these little creatures!"[54]

The use of "rickety" to describe unsound furniture or buildings (or workers) was new in the nineteenth century, but by Tristan's time it appeared frequently in print. This usage rose in tandem with the use of "rickets," "rachitic," and the several terms describing physical manifestations of rickets.[55] Rickety tables and chairs dot the novels of Charles Dickens, and in two instances he applies the adjective to characters, though not in a clinical sense (for example, an exhausted Jerry Cruncher, *A Tale of Two Cities'* cockney "resurrection man," says of himself, "I am as rickety as a hackney-coach.") The word "rickets" itself is missing from Dickens's works, but the consequences of childhood rickets mark the body of one of the most Dickensian of characters, Jack Dawkins, *Oliver Twist's* "Artful Dodger," whom Dickens introduces as "short of his age: with rather bow-legs, and little, sharp, ugly eyes."[56] In his nonfiction, Dickens was more likely to note bowlegged denizens of the streets his alter ego, the Uncommercial Traveler, encountered.[57]

Perhaps the most compelling description of a rickety child in all of Victorian "social problem" literature appeared a generation after Dickens's bowlegged Jack Dawkins. In 1867 social journalist James Greenwood published a feature in the penny press daily *Evening Star*. "Little Bob in Hospital" tells of Greenwood rescuing a child, literally from off the streets of London. Greenwood sets the stage: walking through a neighborhood where the houses are "tall, gaunt, smoke-dried edifices," and the cobble-stone pavement "so favourable to the accumulation and retention of garbage," he comes upon a swarm of children engaged in "vigorous recreation afforded by shuttlecock and battledore." Greenwood's way is blocked by Little Bob, sitting mute and stock still, "his shockingly thin white legs crossed tailor-wise on the chill black pavement, mutely appealing out of his big heavy eyes to somebody to help him."[58]

Greenwood's description of the child's appearance is as accurate a portrait of severe rickets as one would find in a medical
textbook. "He had a monstrously great head, poor child, and a pot-
belly, the two seemingly comprising a burden to which his puny
strength was quite unequal."[59] Greenwood asks one of the children
if the child can walk. "How can he when he's got the rickets?" came
the reply. "Can't you step over him, stoopid?" Greenwood estab-
lishes the facts—Bob has seen no doctors. "What's the use, when
it's rickets?" says the young girl who, it turns out, is paid three
pence a day to watch Bob. "He'll have to grow out of them. That's
wot his young sister died of." Moved to speedy action, within a
week Greenwood has arranged a bed for Little Bob at the Hospital
for Sick Children in Great Ormond Street.[60] Some weeks later,
Greenwood pays a visit to the hospital and finds that the boy, formerly of "towzled-hair and the hollow-eyes so eloquent of pain
o'nights," has been transformed into a "clean, bright, mischievous
little urchin."[61]

Greenwood does not linger in the hospital long enough to
find out what the doctor (Greenwood repeatedly refers to him as
"the magician") actually did to transform Little Bob.[62] Like most
of the social critics of his day, Greenwood did not delve into the
medical aspects of rickets. By contrast, Engels embraced clinical
language, preferring, for example, "rachitis" to "rickets," and he
more explicitly speculated as to the immediate cause of the rickety
children he saw in Manchester. He characterized rickets as a nutritional disease, a consequence of "insufficient bodily nourishment,
during the years of growth and development."[63] But most of the
Victorian social critics did not address rickets with any specificity.
De Tocqueville, Dickens, Tristan, and even Greenwood were less
interested in exploring clinical causes of specific diseases than in
decrying the conditions that produced the general insalubrity of
the nineteenth-century city.

While the Victorian reading public might stumble across the
occasional speculation as to the cause of rickets and its long-term
consequences, medical professionals from Glisson's time forward
consumed a steady stream of case reports, theorizing essays, and

epidemiologic studies. By the last third of the nineteenth century, a few themes emerge from this literature: the unmistakable fact of high prevalence of clinically significant rickets, a clear set of environmental and economic associations, and an energetic (if not always coherent) project to determine the root cause (and hence, potentially, a cure).[64]

That rickets was prevalent in the nineteenth century is easy enough to establish, though with little specificity. Because rickets is seldom fatal, and its signs frequently transitory, its incidence has never been recorded with the rigor or consistency found in reporting other conditions. A survey conducted by the British Medical Association in the 1880s demonstrates this certainty without specificity. The survey asked whether doctors would "be likely to meet with . . . a case a year" of a handful of diseases, including rickets. Organized geographically and plotted in multicolor maps, the results produced strong impressionistic evidence further implicating crowded industrial cities.[65] Other practitioner observations, scattered statistics from pediatric departments and hospitals, and projecting backward from case-finding programs in the early twentieth century, leave a powerful impression of significant prevalence in children of all ages, with rates between 25 and 50 percent for infants and toddlers, and some hospitals on the continent reporting rates as high as 80 percent.[66] It becomes clear that the Parisian "rickets skeleton" in the Warren Museum was only unusual in the severity of its twisted bones. Medical literature had no justification to call rickets the "English Disease." Teutonic children, Gallic, Gaelic, and Norse children in smoky cities all faced much the same risks.

By this time, then, there was wide consensus as to the economic geography of rickets—it was inextricably linked to poverty and squalid housing. What actually caused rickets, however, remained the subject of much speculation and educated guesswork. Sometimes shots in the dark hit the mark: Scottish surgeon William Macewan concluded in 1880 that rickets arose in settings where children were "shut out from the light partly by the height of the houses, partly from the fact that even the sun's rays which

FIG. 1.—Showing degree of prevalence of rickets throughout Great
Britain and Ireland. It is interesting to note that the cross-marking
representing the areas where rickets is very prevalent also closely
corresponds with the areas of greatest prevalence of tuberculosis,
with the great coal-mining and industrial districts, and with the areas
of greatest density of population.

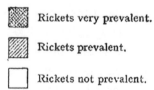

Rickets very prevalent.

Rickets prevalent.

Rickets not prevalent.

Figure 1.2 Geographical distribution of rickets in Great Britain and Ireland, 1889.
J. Lawson Dick, *Rickets: A Study of Economic Conditions and Their Effects on the
Health of the Nation, in Two Parts Combined in One Volume, Freely Illustrated*
(New York: E. B. Treat, 1922), 22.

do manage to struggle through the canopy of smoke which envelops them, are so diluted that they are of comparatively little value."[67] In 1880, Macewan could not know that what linked rickets with "diluted sunshine" was diminished ultraviolet radiation resulting in inadequate production of vitamin D; that connection would not be made for another forty years. Instead, Macewan was drawing on traditional miasmatic theories that found "bad air" itself the cause of all manner of diseases; in his Glasgow, sunlight was prevented from fulfilling its purifying function by the filthy skies and overbuilt streets.

In similar fashion, London physician John Snow, famous for his pioneering epidemiologic detective work in London's 1854 cholera epidemic, drew on medical geography and forensic chemistry to reach his own original conclusion about a cause for England's epidemic of rickets. Looking for a dietary explanation that addressed the higher prevalence in cities, he turned to food adulteration. He published a speculative study arguing that the adulteration of bread with alum was the chief cause of rickets. He recalled from his practice in the North that families there tended to bake their own bread, while in the cities of the South bakers supplied most families with their daily bread. "Now, the bakers," he posited, "all put alum in their bread, whilst this is never done in domestic practice."[68] Snow believed that the alum destroyed the calcium in bakers' bread. He hoped to conduct "an exact numerical investigation," but those plans came to nothing. John Snow's health was failing; chronic illness and his burdensome anesthetic caseload prevented such an undertaking (in fact, Snow died the next year, at the age of forty-five).[69]

"Many things have been put forward at one time or other as essential agents in the production of rickets," pediatrician W. B. Cheadle concluded in 1888, with considerable understatement; among them "congenital syphilis," "inherited constitutional tendency," "defects of hygiene (such as want of light, of fresh air, of cleanliness, of warm clothing, of sufficient food [)]," or dietary "defects."[70] Most of these medical professionals framed rickets as a disease of sunless and miasmatic urban environments, as

Macewan had, or the result of improper diet, whether from food adulteration, as Snow proposed, or a deficiency in dietary fats, as Cheadle and many others believed. The coming understanding of vitamin D and its role in nutrition would reveal that both sides were onto something. In their own time, however, all of the proposed causes for rickets spoke to fundamental social concerns of the Victorian era.

The sources in this chapter remained largely silent on race and ethnicity, factors that would play a perhaps oversized role in the scientific and popular literature since the start of the twentieth century. But well into the nineteenth century, on the eastern shores of the Atlantic and around the Mediterranean, Europeans saw rickets as being shaped by many factors—industrialization and environmental degradation, chiefly, but also by latitude, climate, and other geographic variables, along with differences in lifestyle and social class, with its changing definitions and priorities.[71] All of these played their part in the United States, but there, race in all its manifestations dominated the conversation.

The essential difference between the European and the American experience of rickets can be seen in English literature's most famous ostensibly rickety child, Tiny Tim Crachit: an utterly sympathetic character—the innocent victim of all the evils of the Industrial Revolution.[72] Tiny Tim was a pitiable child, but in his spirit and his flesh a good little English boy, and so the perfect avatar for "The Social Problem" and its cure: if three spirits could transform Ebeneezer Scrooge, who could pay it forward and make Tiny Tim whole again, reformers could work a similar conversion on society at large; a chastened state could end rickets and other diseases of filth and poverty. In the United States, rickety children may have elicited sympathy, but it was a sympathy twisted and stunted by the racial worldview present at the nation's founding.

2

Rickets and Race in the United States, 1492–1900

On the second day of Christopher Columbus's exploration of the New World, he entered into his journal the first written observation—indirect, but telling—of rickets and race in the Americas. On the *first* day of contact, Friday October 12, 1492, the *Santa Maria*'s gunboat had carried the admiral and a small party to the shores of Guanahani, where they were greeted by a host of the island's inquisitive young men.[1] The Europeans spent that first day making small trades with the eager Taino, and—as Columbus's diary entry makes clear—apprizing their fitness to be converted or enslaved. The admiral's observations of the native men's bodies focused on their nudity, their skin color ("neither black nor white"), and their hair ("short and coarse, almost like the hairs of a horse's tail"). As for their frames, Columbus noted that they were "of fair stature and size, with good faces, and well made."

Columbus's diary entry for Saturday, the second day on shore, begins immediately with additional commentary about the bodies of the Taino men. After mentioning once more their "fair stature" and handsomeness, he qualifies that description: "All alike have very straight legs and no belly but are very well formed."[2] There is nothing surprising about a European commenting on the vitality and "natural" handsomeness of Native Americans—such praise would become so common as to approach conquistador boilerplate.[3]

But Columbus's emphasis of the universally straight legs among the natives reads as a clear expression of surprise; it would seem that straight legs were far less common among the Christians aboard the *Santa Maria* than among the naked heathens on the beach.[4] Bowlegged and knock-kneed sailors would have been a common sight in any European port and aboard ships, but they didn't get their rickets on the high seas.[5] Sailors spent most of their time above deck sopping up sunshine, and the sardines and anchovies they ate while at sea were rich in both vitamin D and calcium.

On the other hand, some of the sailors very likely showed signs of scurvy, the vitamin-deficiency disease most closely associated with long ocean voyages. Like most ships of its day, the *Santa Maria* and its two accompanying caravels probably carried little fresh fruit. The resulting lack of vitamin C predisposed sailors to lassitude, severe joint pain, bruising, and open sores. It was not until 1747 that British ship's surgeon James Lind undertook a shipboard clinical trial famously identifying fresh citrus as a miracle cure for this often-fatal sailors' curse.[6] Although there is no direct evidence in his journal, many of Columbus's sailors would probably have been developing scurvy by the time they reached the Bahamas.[7] Severe scurvy leaves marked lesions on the bones, but it does not produce bent limbs. If Columbus saw bowlegs among his crew, they were not from scurvy. Any contrast between the crooked legs of the European crew and the "well made" Taino men was not due to shipboard deprivations, but instead to childhoods spent in dark squalor in European cities.

Because race, not class, came to dominate popular and medical conversations about rickets in modern America, we are primed to share Columbus's surprise: if anyone in this encounter bore signs of rickets, we expect it would be these dark-skinned people living without benefit of Western medicine and Western diet. The underlying reality turns out to have been markedly different from that evoked in so much of American popular and medical literature, different, in fact, from the one I expected to tell—a historical analysis of what *seemed* like an archetypal "racial disease." The historical record frequently maps the burden of disease along racial and

ethnic divides. Within any society, different groups, typically defined by racial or ethnic terms, seem to respond to shared health
threats in very different ways—perhaps partial resistance or distinctive bodily responses to an infectious disease. At one time,
health experts and scholars blithely accepted these associations.
Today, most historians and epidemiologists read those maps
askance. Now they emphasize the social and environmental factors
responsible for these population-wide disparities; they find the
very notion of "race-specific" diseases, susceptibilities, or invul-
nerability problematic. Racism has clear population-wide health
outcomes, they argue; race and ethnicity do not.[8] We should never
accept at face value plantation owners' and racist town boosters'
clearly self-serving conclusions about "race disease." But neither
should we automatically dismiss as mere superstition any patterns we find in the historical record.

Because of the central role skin pigmentation plays in vitamin D metabolism, the history of rickets and vitamin D deficiency
seems custom-tailored to contribute to long-running conversations about race and disease. So I sought traces of the physical
realities of rickets in North America that would mark a clear
history—change over time—in those realities. I probed modern
anthropological literature and historical medical writing in search
of reliably quantifiable data, or at least clear patterns in professionals' records, that spoke to the role of geography, skin color,
and diet in shaping the prevalence of rickets over time, across
latitudes and population densities, and between racially defined
populations. Proof of these changes came up short. I am not convinced that it is impossible to reach a conclusion about these associations, but it seems nearly so. The potential patterns were
continually confounded by evidence of rickets' ubiquity across
racial categories and geography, by the complex interplay of economic realities and cultural practices, and even more by the perennial economics of racism that kept Africans and African Americans
at the low end of the economic bell curve.

What I found instead was clear evidence of rickets' shifting *cultural meanings*. Those meanings were never universally shared: in

America's slave culture, rickets acted as a marker of racial difference for some and proof of slavery's damaging effects for others.
By the end of the nineteenth century, rickets demonstrated, for
progressives and antimoderns, the perils inherent in the modern city, while racists and nativists highlighted the preponder-
ance of rickets among dark-skinned inhabitants as proof of their
racial and cultural inferiority, or at least unsuitability for life in
America's rough-and-tumble industrial democracy. Although this
chapter culminates in the heyday of hard scientific racism, it does
not tackle head-on whether or how rickets should be incorporated
into the historical analysis of "race-specific" diseases. The archival
and archaeological evidence I present does not support such a conclusion. The focus here is the ways contemporary observers made
sense of (and often made fun of) the rickety bodies around them. I
am telling a story about rickets and race; but more important I am
telling a story about stories about rickets and race.[9]

::

As surely as childhood rickets left its mark on the long bones of
its victims, so too was it visible in the streets and public byways
of slave-holding America. The stigmata of rickets were as common
on the bodies of Africans and African Americans as were the scars
left by the lash and would not be easily concealed by garments.[10]
So it is no surprise that enslavers who placed advertisements in
local papers reporting freedom seekers so frequently noted the
bowlegs or knock knees of their absconded human property. In
1826, North Carolina slaveholder M. C. Stephens placed a notice
after Lifford, a skilled blacksmith and carpenter, "stout built, (but
not fleshy,) bow legged, and has a scar on one of his arms near the
wrist," fled from his home in New Bern. Lifford's bowlegs were as
clear an identifier as his stout build, "his complexion black," the
scar on his arm, or his whiskers.[11] Hagai Sexton's "remarkable
bow Legs" did not keep her from repeatedly running from servitude. In her early twenties in 1770, she was indentured as an ap-
prentice to William and Ann Smith of Fredericksburg, Virginia.

A little more than a year later Hagai (or "Hankey" as the Smiths called her) fled."[12] Over the next months, Smith received reports of the young woman (described by him as "a dark Mulatto") as she navigated nearby towns and farms. She was returned to Smith, but disappeared again the next year, and four years later was reported to be passing "for a free woman . . . harboured by a parcel of free Mulattoes."[13]

Whether enslaved or bound to a term of years like Hagai, skilled artisans and house servants appear frequently in runaway ads, raising questions of a connection between rickets and domestic service. Was the number of bowlegged house servants in these ads proportional to rickets survivors in the general population of bound laborers? As tempting as it is to assume that years of indoor servitude might produce rickets in a domestic slave, a more likely scenario has bowlegs or knock knees acquired in childhood impairing an individual's ability to work in the fields, steering them toward domestic service.[14]

The case of Hagai reminds us that not all fugitives were slaves, nor were they all Black, though the runaway advertisements most widely read and anthologized today might support such assumptions. But the more impressionistic search undertaken here, electronically scouring full-text searchable American newspapers for pertinent keywords ("rickety," "bowlegged," "knock-kneed," "bandy-legged") produced more than a few examples from two additional and important categories of freedom seekers: European-born indentured servants and criminals on the lam from justice. In 1769, George Prestman placed an ad in Philadelphia's *Pennsylvania Chronicle* to recover Henry Cartwright, his English-born servant. Cartwright had made off with "a silver watch, two silver spoons and a good beaver hat," and should be easy to spot, being "greatly bow-legged, broad shouldered, with a stoop."[15] Wanted notices for criminal fugitives were just as likely to call attention to the bowed or knock-kneed legs of runaways like the notorious Enoch Calvert, a confessed highwayman, who in 1823 became the first prisoner to escape the newly-built jail in Brentsville, Virginia. The public notices described him as "about 5 feet 10 inches high, sparse made and bow legged, red hair, light coloured eyes, and a florid complexion."[16]

Runaway ads from colonial Massachusetts to late-antebellum New Orleans tell much the same story—the bowlegged freedom seeker would seem to have been ubiquitous, producing a body of historical evidence too scattershot to test theories about changing incidence over time, let alone differential impact across ethnic groups.[17] The ads show that many a field hand, stevedore, and industrial slave bore the skeletal evidence of childhood rickets; but so too did fugitive indentured servants and criminals. And so we turn from the stories of individuals running away on rickets-bowed legs to the stories told through the cold reckoning of hundreds of skeletons disinterred from boneyards. Though their individual life stories remain unknowable, the accumulated data about their dead bones permit statistical analyses by which to infer the lived experience of whole populations.

Guests at the upscale Dominick Hotel (formerly Trump SoHo) on Spring Street in Manhattan probably rest easier *not* knowing they are sleeping atop what was intended to be the final resting place of hundreds of New Yorkers who lived and died almost two hundred years ago. The hotel stands on the site of the Spring Street Presbyterian Church, founded in 1809. Spring Street was famous for its tradition of racial tolerance (it had a decidedly multiracial congregation) and steadfast engagement with abolitionism. In 1963, with its congregation dropping below fifty, the New York City Presbytery closed the church; three years later the building burned and was razed, and its site was turned into a parking lot.[18] It was only in 2006, as construction began on Trump SoHo, that Spring Street Presbyterian's four burial chambers were rediscovered. A backhoe operator unearthed the first cache of bones on Monday, December 11. By Wednesday the medical examiner's office had determined the bones were human, and the city's Department of Buildings issued a stop-work order.[19] Within a week of the site's discovery, a team of archaeologists was racing to exhume and catalog the vaults' contents. Just five weeks after a backhoe first unearthed a human skeleton on the site, the last of the nearly two hundred remains had been removed. Construction resumed at full speed shortly thereafter.[20] Over the next six years, a team of bioarchaeologists at Syracuse University, led by Shannon Novak, studied the remains. In June of 2014, the New York Presbytery took possession of the bones once more and oversaw their reinterment in Brooklyn's historic Green-Wood Cemetery.[21]

Because of the indelible stamp rickets leaves on the bones, burial grounds offer direct historical evidence of its impact—in individuals and potentially in populations. The long bones of roughly one-third of the seventy children buried in the Spring Street Presbyterian burial vaults showed signs of severe rickets; for infants under the age of 1.5 years, the rate exceeded 50 percent.[22] We know little about these children's ethnicity. Only a handful of the Spring Street skeletons could be positively identified, all of whom were European American. Although ample written documentation attests to the fact that African Americans composed a sizable minority of the Spring Street congregation, and anthropologists found "suggestive skeletal indictors of African ancestry in some individuals," the research agenda did not include grouping any of the skeletons based on "racial" skeletal characteristics.[23]

Sometimes there is little doubt as to the racial identity of the skeletal denizens of a particular site, as in the case of New York's African Burial Ground, unearthed in 1991. When excavation began for a new thirty-four-story office building for the federal General Services Administration, grisly finds made it clear that the construction site lay in the middle of a cemetery known to eighteenth-century New Yorkers as the "Negroes Burial Ground." The GSA building is only eleven blocks south of the site of the Spring Street Church, but the African burial ground predated Spring Street by a century or more. The oldest written record identifying the segregated burial ground is a 1712 report of the execution there of participants in a slave rebellion. Black New Yorkers were probably being buried there much earlier, from at least 1697, when Trinity Church, about twelve blocks south, declared its cemetery off limits for "the burial of blacks, Jews and Catholics."[24] At the end of the eighteenth century, when the burial ground was

closed, Africans and African Americans made up roughly 20 percent of Manhattan's population, and the burial ground extended
over five acres of lower Manhattan. The 419 skeletons recovered
from the GSA site represent a tiny fraction of the tens of thousands of Black New Yorkers interred in the greater burial ground.
The vast majority of the grounds' graves were destroyed in the
fast and furious rebuilding of the early 1800s. The tiny portion
of the graveyard unearthed in 1991 survived only because it had
been buried for two hundred years under a twenty-foot layer of
earth deposited when high ground to the north was leveled for
new construction.[25]

A team of anthropologists led by Michael Blakey of Howard
University undertook the task of analyzing the 419 skeletons recovered at the African Burial Ground site. They examined almost
300 for rickets. Almost one in eight of the selected skeletons
showed visibly bowed legs, with twice as many adults as "subadults" having the visible signs.[26] Surprisingly, the children buried
in the African Burial Grounds appear to have far less rickets than
those buried in the vaults of Spring Street Presbyterian, by a factor
of five.[27] This difference flies in the face of modern expectations
about both race and economic status. Spring Street Presbyterian
was set in a mixed working- and middle-class neighborhood, and
was largely European American, while the older burial ground was
by definition racially segregated, filled with the remains of the
city's most downtrodden souls.

Factors other than race clearly shaped this remarkable contrast,
with diet, clothing, and child-rearing practices the likeliest influ-
ences. Place of birth would seem equally important. A significant
portion of Africans and African Americans buried in the African
burial grounds between 1700 and 1800 would have been born
in Africa, on Caribbean plantations, or in the Southern colonies.
Would a childhood spent in a sunny climate reduce the chances
of contracting rickets? Another factor worthy of consideration
is the growing density of settlement in lower Manhattan in the
years the two sites straddle. By the mid-nineteenth century, the
Spring Street neighborhood contained densely packed apartment

buildings, and its increasingly dark streets began approaching the conditions of industrial cities of Europe that were so productive of rickets.

A very different study considered the bones of hundreds of men who were—presumably—never buried.[28] To assess the changing burden of disease of poor Americans born before and after the Civil War, anthropologist Carlina de la Cova studied 651 skeletons from anatomical collections in Cleveland, St. Louis, and Washington, DC. Unlike the unnamed skeletons disinterred in cemetery excavations, the collections de la Cova studied included crucial data about each skeleton, including age at death, race (as reported by the institutions accepting the "donated" cadavers), place of birth, and condition of servitude. These men, roughly equal numbers of African American and European American, were all destitute at death, and most were listed as unskilled laborers. The tale their bones tell is one of uniformly hard lives: four-fifths showed signs of chronic malnutrition, and 4 percent showed skeletal tuberculosis (indicating a very high pulmonary TB rate). As for rickets, 4 percent of the skeletons had bowed legs. This sounds low, but that 4 percent were those whose childhood rickets had never completely healed. Likely many more of these men—an unknowable multiple—had been bowlegged toddlers whose legs straightened spontaneously as they grew. Almost 5 percent (4.8) of the African American men in the collections had bowed legs, compared with 3.1 percent of the European American men. Since the bones were collected over several decades, de la Cova was able to draw some conclusions about changes over time, and while the findings fall in line with recent historical studies that find a decline in health for African Americans after the Civil War, de la Cova found little increase in the prevalence of rickets.[29]

So many variables, so few bones. The anthropological evidence we have does not support the view of rickets as a "racial disease" that emerges in runaway ads and later medical literature. The bone yards and anatomical collections suggest that rickets was only slightly more common in Black populations. We are left with the uncertain sense that geography, built environment, economics, climate, and childrearing practices weighed more powerfully than race in determining the incidence of rickets.

The same evidence does make clear that rickets was a common affliction across race and class that would have been obvious to even casual observers. In 1895 Dr. J. P. West of Bellaire, Ohio, argued that while many of his contemporaries believed it had been "a rare or infrequent disease in this country twenty or thirty years ago," they were ignoring the "evidence perambulating the streets of almost every village." The "rachitic deformities plainly evident to the eye" should convince them "that it was not infrequent even at that date."[30]

::

With this abundant evidence, it followed that descriptions of rickety Americans filled popular periodicals and the daily news, adding the insult of social stigma to the injury of rickets in their skeletons. This literature featured both Black and white rickets survivors, but stark rhetorical differences in these two sets of stories suggest a corresponding contrast in what readers were to make of rickety Black bodies, and how the social onus of disability was joined to assertions of racial inferiority.

In 1899, American humorist Leon Mead published "The Bow-Legged Ghost," a classic example of how rickets in white bodies was understood. Our narrator, suffering from sleeplessness, is on a midnight ramble in the country when he encounters a ghost, "a pudgy figure . . . clad in a white robe." The insomniac strikes up a conversation with the specter who, it turns out, is "the spirit of Peter Simpkins, late of Buffalo, N.Y." Consistent with the habits of the upstanding late nineteenth-century American male he had been, the ghost is on his way to a meeting—"the Annual Convention of Unfortunate Spirits"—a gathering of "second-class" ghosts. The spirit explains he is second-class because he is bowlegged. "People who are badly misshapen, or who have any physical or mental abnormalities, carry them into the spirit world after death." He pulls back his robes to expose his "phosphorescent skeleton." "He

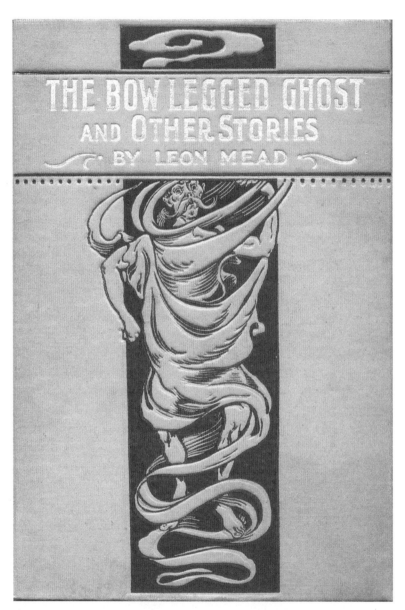

Figure 2.1 Nineteenth-century American popular literature frequently included people bearing visible signs of childhood rickets. Humorist Leon Mead included his short story "The Bow Legged Ghost" in his 1899 collection of his stories.

was the most hopelessly and ridiculously bow-legged individual I had ever seen," the narrator reports, and though tempted to laugh out loud, he is restrained by "the intense glance from his fiery eyes." Though in life, his deformity had not hindered his success in business and society, in the spirit world, he complains that "caste will exclude you from the higher spirit circles, unless you are just right." The ghost is chairing the convention and hopes to be elected the association's president. He is to speak on the topic "Can Second-Class Ghosts Be Happy?" But when next the two meet, the spirit reports that he lost the election—to a Civil War general who "read a paper describing his military achievements at the Battle of Antietam, where both his legs were shot off." The dejected ghost has decided to resign from the association and stay out of politics: "After all, I am only a bow-legged ghost, but to be beaten in a race by a man without any legs at all is a terrible blow to me."[31]

The bowlegged ghost, although in life a prosperous, presumably white, family man, is in death doubly afflicted—doomed for eternity to be "second-class," but also afflicted with a deformity associated with constitutional weakness. References to these skeletal "deformities" in popular culture and the press of the eighteenth and nineteenth centuries suggest that contrary to the ghost's assertion, bowleggedness conveyed second-class status on earth as it did in the afterlife.[32] The runaway ad Polly Lutgert published in the summer of 1800 plays on these associations. The ad proclaimed that her husband, John Charles Lutgert, "did about the first of June last forsake me, and absent himself in a clandestine manner from my board." Polly offered no reward for John's return. On the contrary, she declared "that I do . . . disown and reject him from this time, forever," and requested that readers not "use the least influence in their power to return him to me again." This ad, most likely no more than a newspaperman's humorous bagatelle, is still noteworthy for Polly's physical description of her wayward spouse: "a dirty Dutchman, between five and six feet high, has a round face, snub nose, large mouth and teeth, squint eyed and marked with the smallpox—in addition to this he has long, small legs, knock kneed, cat ham'd, hump'd back, and a fool."[33]

This sort of laundry list of flaws, with a few skeletal references thrown in for good measure, was memetic. For example, under the headline "An Editor Lost," the *Macon Telegraph* sought "a sway-backed, knock-kneed, box-ankled, pigeon-toed, hump-shouldered, cross-eyed dude."[34] In these hyperbolic or cast-off litanies, an imprecise and inconsistent distinction arose around how different rickety defects were moralized. Leon Mead's choice to make his second-class ghost bowlegged and not knock-kneed probably reflected this distinction. By the era of Reconstruction, "knock-kneed" had become the more pejorative—usually associated with cowardice. "Bow-legs come from courage and strong will," insisted Elbert Hubbard, founder of the East Aurora artisan community, citing the example of the child who stands and walks before its time. "Men with knock-knees are weak, vacillating, cowardly. . . . Hounds have knock-knees, bull-dogs are bow-legged. Charles the First was knock-kneed; Cromwell (like Caesar) had bow-legs."[35]

However the popular press characterized those with signs of rickets, bent limbs were ever more visible in print toward the end of the nineteenth century. Changing fashions and lifestyle choices certainly played an important role in this. As Americans took to public beaches and swimming pools, even the modest bathing suits donned by American women revealed more about the legs beneath. One Texas humorist, writing in the persona of "Bill Snort," described eastern beaches in verse, as the place

> Where fat and lean together strive
> To stem the rushing waters.
> And the outside world can view the limbs
> Of fashion's fairest daughters.

Turning his critical gaze upon his own limbs, Snort reported: "I look very picturesque in my bathing suit, but it gives away the fact that I am bow-legged. Both legs are badly warped. A fortune awaits the man who invents a bathing suit that will conceal biased legs."[36]

Other humorists suggested that bowlegs and knock knees were highly prevalent among "fashion's fairest daughters." Satirist

Howard Fielding told of a wager he accepted from a Philadelphia theatrical manager: "I'll bet you $100 that if you visit all the beaches this summer you won't find a woman who looks well in a bathing suit." Fielding accepts the wager and roams from Atlantic City, populated with women from Philadelphia, whom he reported were knock-kneed ("the fault of the cable cars," he speculated), to Newport, where New York's "society girls" display their uniformly bowed legs—due, Fielding argued, to being raised by nurses: "You can't," he mused, "expect a nurse at $12 a month to give a mother's care." "Society children are allowed to walk before they're able, and the tender bones of their little limbs get a twist that money won't remedy."[37]

The theater manager had legs and costumes on his mind because he was staging a new musical revue called *Clothed in a Blush* and wanted his female chorus's costumes as revealing as possible. But given his certainty that "there isn't any such thing as a shapely limb," he would probably opt to keep his chorus in tights—padded tights.[38] A stage costumer told a New York reporter in 1886 that "not more than one-fifth" of his customers went without custom padding, with some of these corrections targeting the aftereffects of childhood rickets. "Some of the prettiest girls," the costumer noted, "will be slightly knock-kneed or bow-legged. We have to straighten them out and produce the fine Venus-like looking forms that you see on the stage."[39]

Did John Wilkes Booth—one of nineteenth-century America's most famous actors, and certainly its most infamous—have good reason to wear such padded tights, to disguise his bowed legs? In the few photographs we have of Booth in form-revealing costumes, his legs appear straight and strong. Still, ample circumstantial documentary evidence suggests that the actor's legs were bowed. As a student at Bel Air Academy, he referred to himself as "Billy Bowlegs," and the schoolmaster recalled Booth as handsome "in face and figure although slightly bowlegged."[40] During his career, the press took little or no note of any such defect. In fact, critics often highlighted his good looks and the extraordinary physicality of his performances. "When he moves," reported a Boston critic, "he does so with that aptness of motion, which forbids the observer to define it . . . He has a physique . . . which is equal to any demands which he need make upon it."[41]

That Booth's father, the great Shakespearean actor Junius Brutus Booth, was famously bowlegged, is immaterial—rickets is not a hereditary condition. But Booth biographers made the connection anyway. George Alfred Townsend, author of the 1865 biography *The Life, Crime, and Capture of John Wilkes Booth*, noted that "his legs were stout and muscular, but inclined to bow like his father's." Turning Booth's legs into a hereditary defect multiplied the assassin's faults, casting him in a drama featuring a visitation of the sins of the father. Some later reports featured his legs in order to recast Booth as a character as misshapen as misguided. "He was also small of stature and bowlegged," one such revised appraisal from the twentieth century put it, "a defect which later in life he always carefully concealed by wearing an ankle-length cloak."[42]

The rise of professional baseball in the second half of the nineteenth century put a lot of American men's legs on display, and sports writers seized on unusual features of players' limbs for "color," as in this typical play-by-play from Texas: "McColley delivered it in time to shut off the little bow-legged duck of a left fielder."[43] The first few generations of professional ball provided a number of noted players with bowlegs, such as Charles Krehmeyer, Frank Leroy, and Gene Rye.[44] But by far the most celebrated was Honus Wagner, the "Flying Dutchman," who had a two-decade-long career in the early twentieth century as shortstop for the Pittsburgh Pirates and was one of the original inductees into National Baseball's Hall of Fame. Wagner's bowed legs were as much a part of his story as his playing: "Nobody ever saw anything graceful or picturesque about Wagner on the diamond," admitted the *New York American* in 1907. "His movements have been likened to the gambols of a caracoling elephant. He is ungainly and so bowlegged that when he runs his limbs seem to be moving in a circle after the fashion of a propeller. But he can run like the wind." The *New York Times* described him as "the bow-legged, crab-walking best ball player in the world." New York Yankees pitcher Lefty Gomez famously said that Honus was "the only shortstop who could tie his shoelaces without bending down."[45] His playing—and his legs—were even immortalized in poetry, by no less a versifier than Ogden Nash, who included him in his 1949 tribute "Line-Up for Yesterday: An ABC of Baseball Immortals":

> W is for Wagner,
> The bowlegged beauty;
> Short was closed to all traffic
> With Honus on duty[46]

In the late nineteenth century—the heyday of "Muscular Christianity" and "Physical Culture"—the American physique was a topic of discussion and scrutiny well beyond the beach, the baseball park, and the playhouse. The daily papers were filled with stories of the New Woman, donning her bloomers, riding her bicycle, performing in ballet recitals. The San Jose *Evening News* might have misjudged the link between exercise and rickets, but they got the dominant tone of the age right when they observed in 1895: "The bicycle is said to make people piegeon-toed [*sic*] and knock-kneed. That settles it. The new woman will have none of it."[47]

Americans' legs became more visible and talked about. Changing fashions, sometimes hand in hand with health reforms, brought the gradual abandonment of bustles and corsets, the subtle lifting of hemlines, the gradual reduction in form-concealing petticoats, and the addition of bathing suits and bloomers. In 1880, editorial writer William Alden warned of the consequences of the fashion for what were known as "narrow," or "hobble" skirts, noting that "a perfect leg is among the rarest works of nature," averring that "the ideal leg is as rare as an ideally clean New-York street." Narrow skirts, according to an unnamed "Baltimore anatomist," were going to produce "startling and painful changes" in the human leg. Alden outlined a laughable Lamarckian progression. "The next generation will be knock-kneed, and its children will add to this peculiarity that of bow-leggedness," producing, in two generations, what would come to be the new "normal" legs, shaped like the familiar typographical sign known as a 'brace.'"[48]

Journalists japed about the more peculiar pronouncements of Physical Culture's enthusiasts, but the new health advocates powerfully influenced public perceptions of healthy lifestyle, fashions, and diet; they also redefined the ideal American body. Bernarr MacFadden, one of Physical Culture's leading gurus, went to great lengths promoting the ideal body, including mounting a series of "Monster Physical Culture Exhibitions" in Madison Square Garden starting in 1903, offering two $1,000 prizes going to "the Most Perfectly Developed Man and the Most Perfectly Developed Woman."[49] The *Times* reported on these exhibitions with the mock-serious tone reserved for such events. One passage focused on the preliminary selection process: "All the bowlegged men and hollow-chested women who sought the chance to exhibit themselves were shown to a lower door . . . One man with legs that were bent under the weight of his thick chest and broad shoulders waited outside of the garden for revenge, but was persuaded by the temptation of free liquor to remove himself."[50] The measure of his frame had been taken, and his imperfect limbs had relegated him to second-class status.

Taken together, all these stories of rickety white Americans, from the prosperous merchant demoted to the status of "second-class ghost" in the spirit world to the strong man rejected from a Physical Culture exhibition in the material world, make two important points about popular perceptions of rickets in late nineteenth-century America. First, they provide ample evidence of bowlegs and knock knees among European Americans of all social classes, a crude set of proxy data suggesting that rickets was hardly confined to poor and dark-skinned children. Second, there is a consistency of tone throughout this literature—a light, often flippant approach papering over the implicit stigmatization and moralism directed at the afflicted. Though a Honus Wagner might so thoroughly overcome his childhood "defect" that sports journalists gushed that he "walks like a crab, plays like an octopus and hits like the devil," Wagner's bowlegs partly defined him; Nash's "bowlegged beauty" was more than a bit ironic.[51] The admiration for those who overcame an affliction, the acknowledgement that many of "fashion's fairest daughters" and leading thespians bore the signs of childhood rickets, or the general lighthearted language in most of these stories do not erase the underlying belief that the stigmata of rickets were more than mere physical attributes; they were at some level a personal failure that warranted derision and moralization. America giggled about the legs but judged the individuals borne on them.

: :

The stories of white ballet dancers, beachgoers, or baseball players with misshapen legs made for interesting press because the handicap was incongruent with the individual's profession or status. These same "defects" in African Americans were treated as defining characteristics to be added to the constellation of racial stereotypes employed against them. It was a casual association dating back to at least the start of the nineteenth century, as in this vignette from a Boston-based humor magazine in 1831: "Thanksgiving Day at length arrived, ushered in by a cloudy sky, and a slight 'flurry of snow,' just deep enough to make the Boston Frog-pond more amusing to the boys, as they glided over the 'kiddledee-benders,' and chased some poor bow-legged n----r from their skating ground."[52] The knock-kneed or bowlegged African American was the butt of many a joke or anecdote. Both conditions could appear in the same story, as in "Perault; or Slaves and Their Masters," from 1843, in which Haman, a domestic slave and trickster figure, composes a song about a young enslaved woman, Sally, courted by two men, "Bow-leg Jim, and knock-knee Joe":

> Our wench Sally she hab two beau,
> Dere's bow-legged Jim and knock-knee Joe.
> To win dis gal to dere embrace,
> Poor Jim and Joe would try a race.
> Jim couldn't run, for tread him toe!

De skin rub off de knees of Joe!
Oh! Sally look sad, and cry at de disgrace,
Dat neider Jim nor Joe were made to run race.[53]

The object of humor here is the same as in quips and anecdotes about bowlegged or knock-kneed whites—the imagined consequences of having legs curved inward or outward at the knee. But Haman's song suggests an additional, crucial point: if these men were not "made to run race," what other tasks might they be unable to perform? To the degree that an enslaved person's "market value" was determined by the amount and the nature of work they were able to do, the aftereffects of rickets could diminish that value considerably.[54]

Examples of rickets undermining the cash value of human chattel pepper the pages of whites' postbellum reminiscences of slavery days. Mark Twain recalled a memorable afternoon from his childhood in Hannibal, Missouri, when he and a buddy were caught stealing peaches from an orchard belonging to "old man Price." Price, as Twain told it, had come to Hannibal from Virginia, with "a raft of bow-legged negroes." The two boys snuck into Price's orchard, gorged themselves on the fruit, and stuffed their pockets with more before they were discovered by one of Price's enslaved orchardists. The boys were too quick for the peach tenders but barely escaped harm when "one of those bow-legged negroes set the dogs on us."[55] Though he does not speculate about whether the man's rickety legs slowed him down, Twain's "raft of bow-legged negroes" implied that Price's enslaved laborers, if not disabled, had diminished capacity as workers or diminished value as chattel. A bowlegged slave's diminished market value provides the resolution to "Mammy's Love-Story," by Julia B. Tenney, published in 1901. It is a bedtime story about Cynthy, the enslaved personal servant to Sally, a planter's daughter. Cynthy's childhood sweetheart had been Tobe, another enslaved house servant, whom she loved for his good heart, not his looks: he was, she recalled, "freckled as a guinea-keat's egg, an' squint-eyed, too," let alone being terribly "short an' bow-legged." Tobe's and Cynthy's hearts were broken when Sally married a young man from Georgia and took Cynthy with her to her new home. Two years later Cynthy and Sally conspire to get her father to allow Tobe to come live in Georgia. The old man offers to sell Tobe to Cynthy, to which she counters that Tobe, a "little bow-legged, squint-eyed, frekle'" man was "no credit to the plantation" since he had "so many fus-class ones dat ain' bandy-legged an' wuthless." The planter relents, proclaiming, "Cynthy, you is won de case. . . . I meks you a present ob Tobe."[56]

Gradually, bowlegs, knock knees, and slew-footedness came to be ever more closely associated with African Americans. The linkage was never exclusive (as our bowlegged ghost makes clear) but for every mention of a "bow-legged Jew" or "knock-kneed Irishman" in the popular press, there appeared a dozen or more Blacks with the stigmata. Perhaps the ugliest example appeared in Hinton Rowan Helper's call for genocide, *Nojoque: A Question for the Continent*, wherein Harper says of the African American, "If he is not slab-sided, knock-kneed, nor bowlegged, is he not (to say the least) spindle-shanked, cock-heeled, or flat-footed?"[57] A famous Democratic cartoon from the presidential campaign season of 1860 featured a noticeably knock-kneed African American as exemplar of the race that the Republicans seemed determined to elevate: "A worthy successor," the cartoon has Abraham Lincoln declare, "to carry out the policy which I shall inaugurate."[58]

In popular literature, rickety legs came close to being treated as a racial trait. Edgar Allan Poe described the enslaved "companion" of the westward pioneer Julius Rodman as displaying "all the peculiar features of his race; the swollen lips, large white protruding eyes, flat nose, long ears, double head, pot-belly, and bow legs."[59] Bowed or knock-kneed legs frequently denoted unsympathetic or unsavory Black characters, from a haughty butler—"a gray-haired, bow-legged old negro"—in a 1909 Broadway play, to an unruly enslaved character Chandler Harris's Uncle Remus describes as "bowlegged en bad-tempered." This story tells how the slaveholder scared this man into behaving by visiting him at night dressed as a ghost—perhaps Harris's attempt at a parable about the origin of

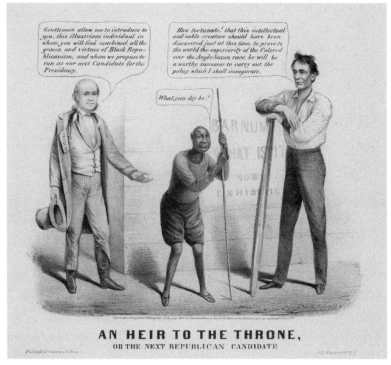

Figure 2.2 Popular media often associated the visible aftereffects of rickets with defects of character, as in this anti-Republican cartoon featuring editor Horace Greeley and presidential candidate Abraham Lincoln extolling the virtues of a man with pronounced knock knees. Currier & Ives lithograph, "An Heir to the Throne, or the Next Republican Candidate," 1860. Artist Louis Maurer adapted this image from a lithograph he produced for P. T. Barnum's "What Is It?" exhibition at his American Museum. Library of Congress.

the Ku Klux Klan.[60] But most of the references were cast off without comment, as with the Massachusetts boys chasing a Black boy off the ice, or the gratuitous description of an enslaved servant as "a remarkably small specimen of a bow-legged 'n_____' " in an article criticizing southern women's use of snuff.[61]

In other instances, the deformity is the center of the story, as in a postbellum reminiscence about a twelve-year-old enslaved child "whose bow-legs had gained for him the elegant sobriquet of 'Pothooks'" being "cajoled and bribed" into dancing for a Northern visitor, his "astonishing and outlandish steps" enhanced by his bowlegs, creating "an intensely comic" performance.[62] One of the most memorable characters in *The Clansman*, Thomas Dixon's 1905 "romantic" history of the KKK, takes advantage of the fully formulated association between rickets and African Americans. Old Aleck, Dixon's braggadocious ex-slave turned legislator, is practically defined by his bowlegs and flat feet. Dixon introduces Aleck as "so bow-legged that his walk was a moving joke." But Dixon took it a step further, treating Aleck's legs as an indicator of his changing mood and status: proudly rising against an insult, "Aleck drew himself to his full height—at least, as full as his bow legs would permit"; believing fortune has smiled on him, "he held himself so proudly that it seemed to straighten some of the crook out of his bow legs"; moments later, when his hopes are dashed, "the bow in his legs seemed to have sprung a sharper curve."[63]

This southern literature featuring rickety African Americans makes it clear that rickets was not confined to the cities of the North. Many of the fugitives from slavery in runaway ads, most of the targets of popular jests, and many who left their bowlegged skeletons, hailed from the countryside of the sunny South. If rickets truly were a "race-specific" disease, this would be expected, since over 90 percent of African Americans lived in the South as late as 1910. On the other hand, evidence of widespread rickets in southern Blacks is at odds with conventional medical wisdom that claimed African Americans "in their natural climate" were not "particularly liable to the disease," that it "is only in the northern climate that their children suffer."[64] Public health orthodoxy defined rickets as a problem of urbanization, congestion, and pollution, but that it fell most heavily on "tropical races" whose migration to northern climates was inevitably unhealthy.[65] Such pronouncements were common well before the modern science directly linking the body's production of vitamin D and the mediating role that dark skin plays in that process. The twentieth century's science of rickets would reinforce the nineteenth century's sense of geographical determinism. The modern reader with some knowledge of how rickets and race are understood might register surprise that the historical record shows such a high incidence of rickets in *southern* Blacks. It is a contradiction that requires explanation. That explanation is fairly plain, if particularly unflattering to the cotton South.

African Americans in the Old South certainly did not lack exposure to ultraviolet light. The vast majority, slave or free, were employed in agriculture, toiling in the hot southern sun. Whether at work or during precious domestic and leisure time, most activities took place outside of the cramped domestic quarters typical of southern slave labor camps. Infants and toddlers, those at greatest risk of rickets, spent most of their days out of doors. Nor were slave children routinely overdressed. Young enslaved children, both boys and girls, commonly wore the skimpy, androgynous "shirt-tail"—a long, loose-fitting smock; only in their teens did girls start wearing long dresses, and boys britches.[66] It was not uncommon, judging by contemporary observers and slave narratives recorded in the 1930s, for enslaved infants and even older children to spend their days naked, or nearly so.[67]

Given this level of exposure to UV radiation, southern Blacks' susceptibility to rickets must be laid on deficiencies in their diet. This conclusion rests on two streams of evidence: the widespread incidence of calcium-deficiency rickets in twenty-first-century Africa and the plentiful evidence of inadequate diets enslaved African American children received.[68] It is difficult if not impossible to describe a universal "plantation diet" in the United States. Regional food supplies and foodways differed dramatically, as did the diets provided by particular enslavers, often supplemented by enslaved laborers themselves through gardening or hunting. Still, we may safely draw two generalizations. First, the typical "slave diet" relied heavily on corn and fat pork—a diet that might supply plenty of cheap calories and iron but was far from adequate for long-term health, failing to provide a full complement of proteins and—crucial for strong bones—deeply deficient in calcium.[69] Second, even if adult field hands were offered ample quantities of this deficient diet, their children were not.

Nursed by overworked and undernourished mothers, southern slave infants were weaned quickly so that their mothers could return to "productive" labor (as well as to "reproductive" labor, which African traditions of extended nursing would have delayed). And the diet to which infants were weaned was scanty and nutrient poor, so poor as to stunt these children's growth. Average heights for enslaved children "fell among or below the poorest populations of developing countries." But unlike populations wracked by ongoing chronic malnutrition, where starvation-stunted children tend to grow into stunted adults, food rationing practices commonly practiced in the plantation South produced, in economist Richard Steckel's phrase, "a peculiar population." Slaveholders nearly starved their enslaved children until their teen years, when they started working the fields. At this point the planters, cynically recalculating caloric input against labor output, increased the hands' food allotments. These field hands underwent dramatic and sustained growth, so that shockingly stunted children grew into adults whose stature matched or exceeded those of working- and middle-class Europeans of the day, seemingly erasing the effects of the near starvation for much of the child's first years.[70] "Seemingly," because despite this impressive catch-up in height, childhood malnutrition left a long shadow in lifelong health outcomes, with the effects of rickets only the most visible.

Slavery's apologists noted only the end result, not the years of deprivation. They did their best to downplay dietary deficiencies in the Old South, pointing instead to the nation's growing slave population in the decades after federal law abolished the international slave trade. That "increase by natural population" was, they argued, "proof of the humanity with which these people are treated." Edwin Clifford Holland's 1822 defense against "the *Calumnies Circulated against the Southern and Western States*" admitted that some areas of the South were unhealthy, but due to environmental factors—the "extreme insalubrity of the air in some portions"—not diet. Careful consideration "to the diseases which afflict our negroes," he promised, "would show, that their food is both more wholesome and more abundant than that of the laboring classes in other countries." As a result of this superior diet, he argued, "dropsies, rickets, scrofula, typhus fever, and the long train of disease which attend upon want and poverty, are far less frequent amongst our slaves, than in England, Scotland and Ireland."[71] Holland's comparative arguments anticipate the later flood of proslavery writers such as George Fitzhugh whose disingenuous comparisons of chattel slavery in the South to "wage slavery" in the industrial North would dominate antebellum defenses against the forces of abolition.[72]

The end of slavery introduces a new set of factors shaping the history of rickets and race in America. By many measures—some visible at the time, some only revealed a century later by historical demographers—African Americans endured a calamitous health crisis after the Civil War. Obviously, emancipation came during a crushing economic and social catastrophe in the South, brought about by the ravages of war and the collapse of the slave economy, leaving many freed people jobless, penniless, and homeless. The health decline began during the war itself: slavery's apologists and abolitionists alike noted the harsh conditions self-emancipated individuals endured and foretold dangerous times ahead for the freedmen. "We shudder to think," wrote the *Richmond Dispatch*, "of the scores of black men, women and children whose miserable deaths are attributable solely to the change in their condition produced by the war." Many expressed concern that emancipation had thrown a generation of bondsmen, wholly unprepared, into a deadly freedom. "Poor dusky children of slavery," lamented Elizabeth Keckley, Mary Todd Lincoln's seamstress, who had purchased her freedom in 1855, "the transition from slavery to freedom was too sudden for you."[73] Sudden or not, the disastrous health outcomes were an integral part of postbellum life for a generation.

Evidence linking the postemancipation health decline to an increase in rickets is inconclusive. A study of height and health data from tens of thousands of prisoners in American state prisons in the nineteenth century supports the now-conventional wisdom: the stature of African American men—a proxy for nutrition, especially dietary calcium and vitamin D—generally increased in the twenty years prior to the Civil War, then declined for a decade or two, consistent with the theory of a postwar health crisis.[74] The prison data also support the conventional wisdom that moving to the city or moving north were unhealthy choices for African Americans, though, of course, staying put on the land in the Jim Crow South obviously presented its own set of risks totally apart from nutrition and health. Whether or not the incidence of rickets actually increased, its presence in the medical literature certainly did. And southern medical professionals increasingly emphasized the determinative role played by race and geography, the factors that would dominate popular and professional understanding of rickets for most of the twentieth century.

Southern doctors acknowledged that rickets had been a visible presence in the plantation South, but they minimized its lasting significance. A year after the Compromise of 1877 ended the Era of Reconstruction and sent the occupying forces of the Union packing, the editors of the *Louisville Medical News* reminisced: "Bowlegged negro babies used to be the rule formerly—frequently the limbs describing almost a complete circle—but they straightened out under the tonic influence of pot-liquor and fat meat." This appraisal seems to corroborate the "peculiar" compensatory growth cliometricians a century later would find.[75]

Increasingly, medical science portrayed rickets as a disease defined by skin color and latitude. "The American negro is very subject to rickets," wrote a Pittsburgh physician in 1895, adding the commonly understood geographic qualification: "It may be said that every northern negro child is more or less rachitic."[76] George Acker, a Washington, DC, physician agreed: "Negroes are almost without exception rachitic." Acker assumed that the "predisposition to rachitis in the colored race would seem to be an acquired one," since, he had come to understand, rickets was rare in Africa. How did African Americans come to be universally rachitic? Although rickets may have been "prevalent even in slave times," emancipation put "the race" at much greater risk. Slaveowners provided their captive labor "food most conductive to their healths." Emancipation, in this view, threw the freedmen and their children to the wolves. From the conventional racist and paternalistic perspective, emancipation left them to their "shiftless improvident ways," and they settled "naturally" to "the lowest stratum of society." Sounding a sentiment common at the century's end, Acker observed that the African American was "undergoing serious physical decay." He concluded with a eugenicist alarm: "This is a serious subject for consideration" he warned, "for we now have a prolific race of unhealthy, degenerated beings who will prove a menace to the interests of the country."[77]

You can well imagine that southerners eager to put Reconstruction behind them and build their New South welcomed this shift in emphasis. The view that childhood rickets was just one of the many deleterious effects of "tropical people" moving to northern climes took the focus off the harsh conditions and paltry diets of slavery days and cleared the stage for New South boosters. The Louisville physician who praised the restorative effects of "pot-liquor and fat meat" in slavery days argued that a new day had come: "How is it under the new regime? Here in Kentucky, we say, genuine rickets—the rickets one sees in the London hospitals . . . does not exist." As vegetable broth and animal fat would straighten a rickety child's bones, restored local control of the Southern economy would restore prosperity and progress—and the good health of the freedmen and their children.[78]

The evidence visible in bones interred in graveyards and the stories pressed into newsprint and vellum make clear just how wrong this moonlit and magnolia-scented vision was—both in its romantic view of the past and its sure forecast for a healthy future. Rickets had been a common childhood malady throughout the nineteenth century, across latitudes and over racial divides. No quantity of potlicker could erase the signs of dietary deficiency that so commonly marked African American bodies, even less so the invisible signs—distorted pelvises and chronic metabolic disorders. Rickets and its aftereffects were among the things southern migrants carried to northern cities in the following decades. But they also posed serious health problems for those who stayed on the land of the semitropical regions of the South, a fact poignantly captured in a 1938 photo by Marion Post Wolcott. It is

December in rural Anson County, North Carolina. Two children approach us over a denuded and deeply rutted field: a toddler, extremely bow-legged, clasps tight the hand of the school-aged girl guiding him over the rough dirt path from the weather-beaten home in the distance. Obviously moved by the toddler's condition, Wolcott included several pictures of him in the portfolio she submitted to the Farm Security Administration.[79]

More conventional evidence of the persistence of rickets in the South abounds. In 1927, flooding of the Red River in southwest Arkansas buried the cemetery of the Cedar Grove Baptist Church under a meter of silt. The cemetery had served the rural settlement's African Americans since the 1870s, but it remained entombed for sixty years after the flood, when construction of a new highway prompted the disinterment of the remains buried there, and anthropologists rushed in to investigate. Their findings confirmed that Cedar Grove's inhabitants had suffered lifelong health effects consistent with the "post-bellum health crisis." Deficiency diseases were a major contributor: one fourth of the individuals buried in Cedar Grove bore the skeletal signs of rickets.[80] In more developed areas of the post-"Redemption" South, rates of rickets were even worse. In 1929, Memphis physician Thomas Mitchell examined a thousand local infants and toddlers—five hundred "white" and five hundred "negro." Almost 90 percent of the Black children showed "definite clinical signs of rickets," a finding aligned with what had by 1930 become the common understanding of rickets, race, and geography. While Mitchell's published findings emphasized the near ubiquity of clinical rickets in Memphis's Black infants, the fact that 50 percent of the white children were rachitic as well confirmed that the problem was far from a "negro disease."[81]

The persistence of rickets in the South, and its persistence across racial lines, requires that we remain circumspect about the conventional view about race and rickets in the Americas. Direct historical evidence—whether anthropological findings or medical reports—frustrated my attempts to plot the incidence of rickets rising with latitude or skin color. The best evidence suggests that

Figure 2.3 Marion Post Wolcott, Negro children and old home on badly eroded land near Wadesboro, North Carolina, 1938. In the four years photographer Marion Post Wolcott travelled the country for the US Farm Security Administration, she took over nine thousand photographs documenting rural Americans struggling during the Depression. Library of Congress, Prints & Photographs Division, Farm Security Administration/Office of War Information Black-and-White Negatives.

rickets was common among Black and white Americans, North
and South. But the story of rickets in America's popular memory
is heavily skewed toward a racial explanation. It matters little
whether this racial association reflected reality, because the rhetoric of degeneration of the bones lined up with the general deep-set
rhetoric of racial disparagement.

: :

The lessons of this and the previous chapter should erase any in-
congruity in what Columbus reported from that beach. Why did
the Taino have such straight legs? Plenty of sunshine and a diet
that provided ample calcium and phosphorus. Why did the legs
of Columbus's sailors suffer by comparison? Because many had
endured malnutrition as children, crowded in dense and sun-
choked hovels in European port cities. We might be tempted to
see these factors—structural inequality with food insecurity and
unhealthy urban conditions—as additional ironic elements in the
Columbian Exchange. Columbus's encounter projects the reverse
image of race and rickets that modern rickets research has led
Americans to expect—dark people with straight legs and white
people bowlegged and knock-kneed. That reversal changes nothing. The point of the American story of rickets was, and largely
remains, that race was the crucial variable in explaining its history
and distribution—not class, as in Europe.

Four hundred years later, America faced a rising tide of new
immigrants. The cities of 1892 in which these masses of new
strangers huddled were themselves rapidly changing. With their
increased density and the thickening atmosphere of light-trapping
pollution, American cities came to resemble Dickens's London,
Engels's Birmingham, or Balzac's Paris—cities starved for light,
filled with populations as starved for nutrients as those enslaved
children in the American South. The emerging cultural image of
rickets as a racial disease further stigmatized many of these new
urban migrants, casting them as less sympathetic than degraded.
The dramatic rise in rickets at this time provided an urgent subject

for scientific study, discovery, and medical intervention. But with the particular mix of scientific energy and demographic and social foment in the United States at the start of the twentieth century, everything was in place for a full-blown rickets "epidemic" of a particularly modern American sort—highly visible, perceived chiefly as an urban problem, and presumably worst among "exotic" racial and ethnic populations. The cities were straining under unprecedented population growth—growth driven, it seemed to the dominant society, by the influx of darker-skinned populations. Rickets, in this setting, was the perfect "racial" disease.

Diet or Light?

The Science of Rickets I

On the first day of spring in 1934, Philadelphia physician Joseph Stokes wrote to the eminent Johns Hopkins pediatrician Edwards A. Park for a favor. "We have at the present time," Stokes explained, "an extremely interesting group of twenty-four colored infants on a separate ward, whom we are studying in relation to irradiated and non-irradiated evaporated milk," and Stokes hoped Park would pay a visit to Children's Hospital to consult on x-rays for the study.[1] The "interesting group" were two-thirds the way through four months of sunless confinement, "participants" in a clinical trial testing whether vitamin D–fortified evaporated milk would prevent or cure rickets. Stokes and his team had selected that "separate ward" because it could "be well ventilated without opening windows so that the children receive abundant fresh air without any exposure to direct sunlight or sky shine," thus "minimizing any slight antirachitic effect of the sunlight."[2] The goal of the experiment was to produce rickets in the toddlers in order to test whether evaporated milk made an effective delivery device for industrially produced vitamin D.[3]

A major research hospital depriving infants in their care of light? Surely this flew in the face of a central tenet of early twentieth-century modern medicine. Few things were more certain in American medical thinking a century ago than the healing power of fresh air and sunshine. Heliotherapy, the practice of exploiting the sun's power to heal, was built into therapeutic regimens and into the very architecture of hospitals themselves. American children's hospitals enthusiastically embraced heliotherapy, including Children's Hospital of Philadelphia: even on the coldest days of winter, young patients spent hours on rooftop patios or in makeshift solariums, soaking up the low winter sun's feeble rays. Some patients were recuperating from operations, others fighting off infections or seeking relief from tuberculosis. And among these young sunbathing patients were a few being cured of their rickets.

Stokes's experiment was hardly groundbreaking. It was one of dozens of human-subject studies rickets researchers mounted in American hospitals, clinics, and community health centers in the first third of the twentieth century. In fact, it came toward the end of two decades of fruitful research into the basic science of nutrition in general, and rickets in particular: between 1917 and the 1930s researchers proved that both cod liver oil *and* sunlight prevented and cured rickets, isolated the specific band of light in sunshine that did that work, identified the specific antirachitic component in cod liver oil, named it vitamin D, and developed the technology to produce that "sunshine vitamin" for commercial distribution. On a different front, orthopedists developed new techniques to correct the aftereffects of severe rickets. In short, researchers on all these fronts invented the modern science of rickets.

The puzzle at the heart of this research was the relationship between rickets, sunshine, and diet. The medical field was split between those who with good reason claimed either light or cod liver oil was a "specific" for rickets—that rare item before the twentieth century's "magic bullets" and miracle cures—a particular remedy for a particular disease (other examples available then included quinine and digitalis). But how could such disparate agents have such similar effects? Sunlight and fish oil—environment and aliment—for centuries, both figured prominently in explanations for why some infants' bones went soft.

: :

Back in the seventeenth century, the first published treatises on rickets by Daniel Whistler and Francis Glisson floated both dietary and environmental (or lifestyle) causes. Whistler found different causes for the children of the rich and of the poor: poor children suffered from "mistakes in diet," but also "lack of fires and a daily environment of dung and dirt," while the rickety children of the wealthy owed their plight to "intemperance of the parents and the fact that the infants are entrusted to the care of hired wet nurses."[4] Glisson's patients were the children of the wealthy, and he laid much of the blame at the feet of their parents, whose diet he found "over-moist and full," and their lifestyle leading to a "soft, loose, and effeminate constitution." Glisson railed against the lack of "strong and Masculine exercises," their love of "Comedies and other Plays," "assiduous reading of Fables, and Romances," and "a loos expence of time in Carding and Dicing."[5] As for environment, Glisson struggled to portray rickets as a "Country Diseas," but try as he might to blame the environmental particulars, asserting that "an excess of cold and moisture may be imputed as a fault to *England*," he had to admit that other countries "far exceeding *England* both in cold and moisture," had little rickets.[6]

The split focus on environment and diet continued through the nineteenth century, as did the question of whether poverty or privilege posed the greater risk. In his popular 1834 *Home Book of Health and Medicine*, American anatomist William Horner asserted that the children most at risk were "those who live in moist and damp places, who are poorly fed, and who are not kept cleanly." He embraced environmental, dietary, and medicinal cures: for settings, he counseled "warm or sea bathing, the pure air of the country, removal from damp and moist places"; dietary cures, including rhubarb and iron; and remedies from nineteenth-century "heroic" medicine's toolbox: "a little calomel [a mercury-based concoction], and an occasional emetic."[7]

Doctors sometimes expressed mild frustration that rickets, a specific disease, had no specific cause. "Numerous theories have

been offered as to its etiology," declared Maryland physician
C. W. Mills in 1894. But those theories did not amount to a single
explanation, leading him to conclude that "there is probably nothing specific in the nature of rickets—that it is a perversion of the
normal metabolism of the body, and that any cause of general malnutrition may give rise to it." He suspected the "artificial" feeding
of babies and speculated that while "the breast-fed offspring of
the poor woman is less apt to suffer from rickets than the artificially fed child of the society woman," the "gross deformities" of
rickets were less visible in the children of the rich "because when
they begin to show themselves they are properly treated." Mills
recommended five treatments, including the two that would prove
to be effective: the rickety child "should be kept in the sunlight as
much as possible. Phosphorus, cod-liver oil, iron and arsenic are of
value medicinally."[8]

The uncertainty about rickets' cause embroiled it in a wide
array of health issues of the late nineteenth century, especially
respiratory diseases. Severe rickets can contort an infant's growing rib cage, increasing the risk of serious, even fatal, respiratory
complications, including the century's top killers, tuberculosis
and pneumonia—the "white plague," and "captain of the men of
death" their respective monikers.[9] In an age predisposed to look to
the lungs as a seat of illness, and witnessing a rising tide of rickets,
some physicians were bound to reverse cause and effect. British
physician Robert James Lee argued that "the precedent cause . . .
of rickets is some form of pulmonary disturbance, generally indicated by the term 'bronchitis.'" Rickets, like pulmonary disease,
Lee argued, fell harder among poor children because of economics,
not because it was "transmitted from parents to their offspring."[10]
Some observers did argue for genetic causes, perhaps none in a
more singular way than British physician Arabella Kenealy, who
argued that modern women's obsession with exercise enfeebled
their offspring. The healthiest children had mothers "whom the
world would call delicate . . . , but delicate only in the absence of
that robustness which is degeneration from the womanly type."
She contrasted the "straight, beautiful limbs" of the children of

such a "delicate" mother with "the puny sickliness, the spectacled, knock-kneed physiques" of those of the "Amazon mother."[11]

If experts remained confused over rickets' exact cause, its increase in the last decades of the nineteenth century was obvious and alarming. Though he argued that rickets was "still rare among Anglo-Americans," a physician in Buffalo noted its sharp rise among the children of Italian immigrants. "With the first generation of Neapolitans born in American cities, all of the causes of rickets—improper food, bad air, depression caused by infantile maladies—are intensified and accentuated."[12] This belief inspired further and more intensive study. And while the two dominant streams of inquiry, environment and diet, persisted, by the early twentieth century, advocates for each had focused on their own particular subtopic. Investigators who emphasized diet more or less stopped speculating about such broad concerns as mere caloric intake or the mix of proteins and carbohydrates. Instead, years before the fat-soluble vitamins were discovered and vitamin D was identified, they turned to the role of insufficient fats in the diet. And those seeking environmental causes turned their attention to light—quite apart from other environmental issues such as miasmas, dampness, temperature, or even toxicants—the simple quantity and quality of light.

Theobald Adrian Palm was uniquely qualified to explore how sunlight, environment, heredity, poverty, and climate shaped rickets in the nineteenth century. Palm was born in Colombo, Ceylon, in 1848, to Scottish Presbyterian missionaries. In his midtwenties, having completed his medical training in Edinburgh, and newly married, he moved to Japan to take up his father's and grandfather's calling as a medical missionary. In Tokyo he and his wife Mary, now expecting their first child, studied Japanese and planned their mission in the hinterlands. But Mary and the child both died in childbirth, leaving Theobald alone to head out to the small Japanese port town of Niigata, where for eleven years he preached the Gospel—and the "gospel of health." He established a dispensary, known as "Palm Hospital," where he treated over forty thousand patients and trained local doctors in Western medicine.

In 1879 he married Isabelle Collas, the daughter of a missionary in Hokkaido. His ministry to health was more successful than his service to the church in the staunchly Buddhist region.[13] In 1883 he spoke in Osaka to a conference of interdenominational medical missionaries, telling them that Japan's medical services were now sufficiently modernized that Western missionaries were no longer needed. He and Isabelle returned to Great Britain the next year.[14]

Practicing in Birkenhead, across the Mersey from Liverpool, he noted the "lamentable frequency" of rickets "among the poor children of the large centres of population in England and Scotland," in direct contrast with "the absence of rickets" in Japan." Palm hypothesized about a "want of light" being the chief cause. So, taking as inspiration two recent published surveys—one charting the distribution of rickets in the UK, and another from hospitals in Europe and Asia—he conducted a survey of his own. He polled former colleagues in medical missions throughout Asia about the prevalence of rickets, the "habits of the people," sanitary conditions, and climate in their areas.[15] The results lent strong support to Palm's thesis about light's role: climatic factors such as rainfall, altitude, and topography; economic factors such as urban blight, poverty, malnutrition—none produced rickets in these Asian lands. One of his informants, William Huntly, reporting from the dry and mountainous northwest region of Rajputana, provided perhaps the clearest affirmation. Huntley was from Glasgow, and so, like Palm, was primed to look for rickets. But in treating ten thousand children over two years in Rajputana, only once had he "occasion to write the term rickets opposite the name of any child." General conditions were not better than in Scotland; in fact, he judged them as uniformly filthy. Indian houses, he reported, had worse ventilation and were smaller than the homes of the poor he remembered from Glasgow. What, then, explained the freedom from rickets? Huntly asserted: "The grand counterbalancing fact which stands out is the sunlight." "As the child steps over the door of the low hut, he passes directly into sunlight," a direct contrast with the houses in large British cities, commonly multistory and built around small courtyards into which "the sun never enters."[16]

Palm insisted that sunshine explained far better than other factors the distribution of rickets, both across the globe and within any particular land. To those who attributed rickets to venereal disease, Palm countered that "there is probably no country where syphilis abounds more than in Japan, and yet rickets is extremely rare." Nor was poverty a reliable predictor: hygienic conditions and diet—two signal markers of poverty—produced surprising results. Sanitation in Asian cities, Palm argued, was uniformly inferior to that found even in the slums of Great Britain. And diet? "The wives and children of the working classes of Britain are better fed than the teeming populations of China and India, who are strangers to rickets." As for differences within Britain, the upending of expectation continued, as "the English town-bred rickety child and its mother are, as a rule, as well or better fed than the healthy child and wife of the agricultural labourer."[17] No, Palm found "the most salient fact" predicting freedom from rickets was "the abundant sunshine and clear sky." Admitting that Britain was by no measure a sunny land, he emphasized the built environment's role in producing the sunless nursery for rickets. The cities were "under a perennial pall of smoke," and rickety children would be "found in abundance" where the "exclusion of sunlight is at its worst": "in the narrow alleys, the haunts and play-grounds of the children of the poor."[18]

If want of light were the chief culprit, a better understanding of "the physiological and therapeutic actions of sunlight" required more study into what Palm called "the Chemistry of Light." There was something specific about the light in different settings—different altitudes, latitudes, natural differences in atmosphere shaped by climatic and geographic conditions—or urban and industrial factors. Noting that the study of light's relation to inorganic chemistry had led to many discoveries, and "the marvels of photography," Palm predicted that the "effect of light in organic chemistry," and in "animal nutrition" specifically, "must be still more full of interesting facts."[19]

Many of those "interesting facts" would come from fractionating the light spectrum into its component parts and exploring wavelengths beyond the visible—a project already almost a century old in Palm's day. By 1801 William Herschel in England and Johann Ritter in Germany had discovered unique light energy both below and above the visible spectrum: Herschel found the intense heat of infrared at one end of the rainbow, in the dark region just below red; Ritter discovered the chemically active band of light just above violet—ultraviolet light. UV light would, of course, be the key to understanding the physiology of rickets and the chief means for fortifying foodstuffs with vitamin D. The "chemistry of light" eventually revealed the entire electromagnetic spectrum, from the low frequency of radio waves far below red, to the rapidly oscillating gamma rays far beyond violet. In the following years, dozens of scientists and tinkerers invented the devices to detect—and produce—electromagnetic waves within the spectrum of visible light and well beyond in both directions. In Palm's time, though more or less irrelevant to his study, came Wilhelm Röntgen's 1895 discovery of x-rays. Röntgen's rays would quickly become an important tool in the study of rickets, just as the electric lamps that produced powerful beams of the ultraviolet light Ritter had discovered and named would define its treatment.[20]

Of course, nineteenth-century health experts did not wait for science to dissect the rainbow before they put sunlight to work. Their long-standing medical cosmology led them to expect disease to thrive in the foul smells, the stale and close atmosphere, and the darkness so common in modern cities; they took as a given the healthful properties of sunshine and assumed a causal relationship between sunny housing and healthy populations.[21] Similarly, many health professionals embraced cod liver oil as a "specific" for rickets long before science could explain why it worked. In the next chapter, I will explore how cod liver oil came to be every mother's go-to tonic, every child's torment, and a great hope of physicians and public health activists struggling to push back against endemic rickets. Cod liver oil's power against rickets was already well established (if not explainable) by the late nineteenth century, so it naturally played an important role in the studies to understand, prevent, and cure rickets. Those studies sought answers to two related questions. First, given the demonstrated record of cod liver oil's effectiveness, how should it best be used as a public health measure? And second, just what was it in cod liver oil that helped build or repair bones?

: :

Both questions were all but answered by 1934 when nurses at Children's Hospital laid those two dozen infants in their dimly lit bassinets. The road to that particular experiment—with its narrow goals, its assurance of its place in a chain of discovery, and its blithe indifference to ethical concerns—was paved by decades of scientific and technological developments advanced through newly formulated biomedical models for research; it was a road to a promised land of magic-bullet medicine and science-based public health. Studying the physics of light had opened a wide spectrum of visible and invisible energy to practical applications and further study. The American food sciences, from their new homes in land-grant universities from Ithaca to Madison, explored the metabolic processes that turned crops into energy for growth, strength, and health, preparing the ground for serious study of vitamins and other micronutrients.[22]

Big economic interests promoted the project to study food scientifically. Large-scale agriculture and food industry interests gained influence at state land-grant universities, whose research agendas reflected the not-always-invisible hand of agribusiness in shaping their priorities. Chemistry, increasingly wedded to the economic power of industries hungry for its fruits, had more and more to say about nutrition and health. Food science and chemistry joined forces on several fronts: analyzing the components of food and applying the principles of chemistry (or merely applying chemicals) to increase yields or enhance food preservation. Long before the end of the nineteenth century, the seeds of our modern discomfort with chemistry's potentially malign role in foods had already been sown in regular scandals involving deliberate adulteration or chemically engineered foods: from the benign—margarine as artificial butter, say—to the deadly—chrome yellow lead mixed into bread as "egg substitute."[23]

As important as the rise of modern food science, and more directly related to finding a cure for rickets, was the modern biomedical model for studying disease. The familiar players in this project—Pasteur, Koch, Lister—were, in the parlance of later hagiographies, "microbe hunters" seeking the seeds of nature's deadliest diseases.[24] The search for dietary illnesses differed in one fundamental way from microbe hunting. Traditional medicine sought a dangerous, often invisible agent that caused disease: miasmas and effluvia in earlier years, germs (or toxicants in occupational settings) later. Microbe hunters sought something that when *introduced* to a healthy body sickened it. And in fact, some pursued an infectious source of rickets.[25] Nutritional diseases, on the other hand, were produced by an *absence*—some unknown, invisible factor that when *withheld* from a healthy body sickened it.[26] Christiaan Eijkman's research on beriberi and Joseph Goldberger's discovery of the cause of pellagra are the most dramatic examples from the late nineteenth and early twentieth centuries.[27]

Though the study of nutritional disorders took a distant back seat to microbiology, the same revolution in biomedical research methods defined both projects. Previous generations understood rickets as an indistinct malady—a constitutional disorder, a racially embedded environmental health problem, or merely a side effect of a generally unhealthy diet. Nutritional researchers in the shadow of Koch and Pasteur sought *the* cause of the distinct condition called rickets and the magic bullet to defeat it. Just as Koch isolated and incubated the tubercle bacillus found in sick bunnies, then infected healthy rabbits with his cultured germs, and in so doing turned *consumption* or *phthisis* into "tuberculosis," by the early twentieth century everything was in place to find *the* cause of rickets and so to cure it: well-funded and highly motivated food science centers in university research centers, medical and public health professionals eager to understand a serious health issue of growing concern, and the new methods of inquiry associated with modern biomedical research.

One of the strongest parallels between the "microbe hunters" and the nutritional scientists of the day was their reliance upon both animal and human subject testing.[28] In the case of rickets, these tests can be arrayed along a continuum (or descending spiral). At one end were those more-or-less benign "natural experiments" in which researchers introduced a therapeutic measure in a setting where rickets was already a debilitating problem. On the other were those experiments performed on healthy institutionalized infants.[29]

London's Regents Zoo was the site of an influential natural rickets experiment in the 1880s.[30] A lioness there had a decade-long history of lactation problems, unable in litter after litter to produce enough milk for her cubs. All the cubs in nineteen of twenty consecutive litters died from severe rickets. At the time, a young surgeon named John Bland-Sutton worked part time as one of the zoo's prosectors, performing dissections and conducting necropsies—both in service to the zoo and to advance his own research interests. He had taken this position several years earlier, while a student at nearby Middlesex Hospital. Even then, Bland-Sutton was an experienced anatomist, having earned his medical school tuition by teaching dissection and anatomy classes to medical students at the somewhat disreputable London School of Anatomy and Physiology, run from a tin shed in an abandoned churchyard.[31] Both his deftness with the knife and his abiding interest in comparative anatomy arose from long experience working in his father's small in-town farm, fattening and slaughtering stock for the London meat market. In his midthirties in 1888, he had risen to the position of Hunterian Professor at the Royal College and assistant surgeon at Middlesex Hospital. All the while he had kept his part-time post at the Regents Zoo, possibly in part because it aligned with a lifelong professional interest in comparative anatomy.

Bland-Sutton had been studying rickets in the zoo's animals, and in 1888 published an overview of "Rickets in Monkeys, Lions, Bears and Birds."[32] It remains unclear how or exactly when Bland-Sutton came to be consulted in the case of the rickety lions, but upon learning of the ongoing problems, and suspecting a lack of dietary fat was at fault, the young prosector convinced the lion keepers to change the rickety cubs' diet, from lean horsemeat to a combination of goat meat, ground goat bones, and milk. And, for reasons Bland-Sutton did not record, he added cod liver oil to their diet.[33] Bland-Sutton and the zookeepers made no changes in how the lions were housed. "They were kept in the same dens, with the same amount of air, of light, and of warmth as before."[34] Three months later "all signs of rickets had disappeared," and when eighteen months old, the cubs were reported "perfectly strong and healthy and well-developed," a claim vitamin D researcher Elmer McCollum repeated decades later. "For the first time lions born in captivity were reared without deformities so severe that they were unfit for exhibition."[35]

Initially, Bland-Sutton's experiments had little impact. The young surgeon continued with his comparative anatomy studies at the zoo but focused on his surgical practice, becoming a prominent abdominal and obstetric surgeon. Apart from a brief mention in his 1888 study in comparative anatomy, he never published the case of the rickety lions.[36] Still, a generation later, when researchers took up in earnest the study of rickets, Bland-Sutton's intervention with those London lions would be cited as an influential experiment establishing an "animal model for rickets," and mentioned as inspirational by later generations of nutrition researchers.

Progress in the years immediately following Bland-Sutton's experiment is best characterized as plodding and unfocused. But in the second decade of the twentieth century, the harrowing burden of a full-blown rickets epidemic reignited interest. Surveying the field in 1922, one of the leading researchers studying rickets, Alfred Hess, crowed that interest in rickets was so intense that "advance in our knowledge promises to be greater during this decade than throughout the preceding 250 years" and that "we no longer are groping."[37] Not groping, but still deeply conflicted about whether rickets was essentially an environmental or a dietary disease. Within the next few years, that acute interest and intense activity would lead to finding the active ingredient in cod liver oil and solving the riddle of how either light or diet could cause or cure rickets.

"Both sides," whether focusing on light or nutrition, shared a set of fundamental problems: finding the cause of rickets and identifying and testing interventions to prevent or cure it. And in studies undertaken in the 1910s and 1920s, both sides followed two starkly different paths simultaneously: some studied children in cities, the dark streets and endemic poverty providing ample natural "clinical material" for natural experiments or community studies; others studied animals in laboratories, concocting controlled studies to provoke rickets in healthy subjects, via diet or environment, to study both the course of the disease and its cure. Eventually those methodological paths got crossed in dozens of studies that applied the veterinary experimental model to human subjects—healthy children in clinical, nontherapeutic studies.

Vitamin D's very name suggests that its discovery falls well along in a larger history of research into the so-called accessory food factors, substances other than macronutrients such as fats, proteins, and carbohydrates. Each of these "discovery stories" includes three roughly similar achievements: the correct linking of a food source for the given micronutrient and the disease resulting from a diet deficient in it; the isolation of the micronutrient; and, finally, the chemical identification of that substance. Those timelines seldom coincide. Take vitamin C: as early as the sixteenth century, sea captains provided citrus or sauerkraut for their sailors with the direct intent of preventing scurvy. When Casimir Funk popularized the misnomer of "vital amines," the reality of an antiscurvy micronutrient and its sources were so clear as to earn its "C," but ascorbic acid itself was not discovered until the 1930s.

Vitamin D's trajectory is similarly bumpy. As in the case of vitamin C, vitamin D had, in cod liver oil, its "lime juice," a long-famous (if scientifically contested) dietary source. As for its later discovery, hence the letter "D," while the micronutrient that prevented rickets was the fourth vitamin to be added to the alphabetic panoply, it was initially thought to be one with the first, vitamin A, because vitamins A and D (as well as E, it would turn out) were often commingled indistinguishably in the oils the vitamin hunters tested. Conveniently, because of this confused origin story, it makes good sense to start at A, with Elmer McCollum and his research partner Marguerite Davis.

McCollum had recently received his PhD in organic chemistry from Yale when he took a position at the University of Wisconsin's Department of Agricultural Chemistry in 1907. He joined a team of researchers at UW's Experiment Station developing a scientific diet for cattle. The "single-grain" experiment tested three diets, based in wheat, corn, or oats, each fortified to provide the hypothetical ideal balance of carbohydrate, protein, and fat. McCollum was responsible for chemical analysis of the rations going into the cattle and the dung coming out. The four-year study found that cows fed a corn-based diet thrived, while those fed a similarly fortified wheat diet suffered from stunted growth, miscarriages, and blindness. Suspecting that corn contained some accessory food factor missing in wheat, the team enlisted McCollum to find it. McCollum convinced the administration, against the college's strong preference for studying farm animals, to allow him to conduct experiments on that traditional barnyard foe, rats, whose rapid growth and reproductive cycle would speed up the process of testing diets. McCollum is credited with establishing the first colony of white lab rats for nutritional studies. Strictly speaking, though, his rats were not like today's genetically engineered identical specimens. In 1908, lab rats were not readily acquired. At first McCollum captured rats in the college's stables, but they were too ornery, so he bought a dozen albino rats from a Chicago pet store.[38]

Over the next five years, McCollum and his graduate assistant and research partner, Marguerite Davis, experimented with their rats, eventually identifying a factor found in butter fat and cod liver oil necessary for growth, and which prevented the blinding eye disease xerophthalmia. This fat-soluble substance would eventually be found to comprise three separate components we now know as vitamins A, D, and E. McCollum and Davis reported their findings on the fat-soluble factor in 1913, only weeks ahead of a parallel study at Yale by Thomas Osborne and Lafayette Mendel.[39]

Chroniclers of vitamin D's discovery usually credit British physiologist Edward Mellanby with the next important step, portray-
ing him as if he were the Robert Koch of rickets. And indeed, the
famous studies he conducted, some in partnership with physiologist May Mellanby (née Tweedy; the two physiologists married in
1914), followed the familiar path of "microbe hunters": provok-
ing rickets in healthy organisms in order to identify the agent
responsible, then testing by withdrawing and reintroducing the
agent to cure or cause rickets.[40] In 1913, Edward joined the faculty of London's King's College for Women, as lecturer in physiology. Up till then he had not studied nutrition, focusing instead
on the biochemistry of muscles. Regardless, the Medical Research
Committee, newly established by the British government, assigned the young hire to try to find the cause of rickets.

Initially, Mellanby did not investigate the possible role of accessory food factors.[41] Instead, he spent the first two years looking for a biochemical cause for rickets, manipulating the diet of
puppies to raise and lower their metabolisms. For the first years,
the puppies were not allowed to exercise and were housed in a
sunless basement at the college in Kensington. Crowded conditions related to the war led him to move his kennels to the sunless
quarters of the Field Laboratories of Cambridge University.[42] As
Mellanby recalled, his charge was to link oxidation and rickets. He
confided to a colleague his frustration at this dead end. "I wasted
a year or probably two years doing all kinds of funny things in
order to produce rickets in dogs by interfering with conditions of
oxidation."[43]

One of those "funny things" was to vary the protein content his
puppies received, since high-protein foods raised oxidation rates.
The puppies fed a high-protein diet developed rickets, prompting
a new set of experiments focused on accessory food factors. He
preferred dogs for rickets experiments because puppies were readily made rachitic. Once a colleague worried why Mellanby had not
always used dogs of the same family; he agreed that "it would be
easier . . . but that condition I have not been able to secure. I have
had to work on any puppy I could get hold of."[44] He put the puppies

on a diet of oatmeal, rice, and milk. After three months the puppies had developed all the signs of rickets. Puppies fed skim milk
developed the symptoms sooner, pointing to an accessory factor
present in the fat, which he presumed would be McCollum's fat-
soluble A.[45] Tests of his "rachitic diet" fortified with various fats
and oils demonstrated wide variability in the oils' antirachitic
qualities, with cod liver oil head and shoulders above the rest.[46] A
deficiency of *something* found in these oils would lead to rickets.
"It seems clear," he announced, "that rickets is a deficiency disease
of the type of scurvy and beri-beri," and that the key substance
was "either fat-soluble A, or has a somewhat similar distribution
to fat-soluble A."[47] It took a few more years to demonstrate that
vitamin A itself had no effect on rickets, and that Mellanby's "key
substance" was a unique chemical. In 1922 McCollum, by then at
Johns Hopkins University, developed a method for isolating the vitamin D: oxygenating cod liver oil destroyed its ability to promote
growth and prevent xeropthalmia—two of three traits McCollum
had identified for "fat soluble factor A," but it retained its effectiveness as an antirachitic. By then the ABCs of naming vitamins
was gaining acceptance: A for the fat-soluble factor McCollum
identified earlier, B and C the water-soluble factors responsible for
preventing beriberi and scurvy, respectively. McCollum and his
team named the antirachitic fat-soluble factor they had isolated
"vitamin D."[48]

Finding and naming a micronutrient that could prevent or cure
rickets did not settle the matter, however. Strong expert voices
persisted in support of "hygienic"—or environmental—causes,
enumerated by John Lawson Dick, the British physician we met
in chapter 1 arguing for the absence of rickets in ancient days:
"(1) the breathing of a vitiated atmosphere in close and confined
dwellings; (2) the exclusion of sunlight; (2) the lack of opportunity of exercise; (4) damp climates and long winters, notably in
the colder northern temperate zones."[49] Dick assured his readers
that the evidence for rickets as a dietary deficiency "is still entirely
wanting."[50] Why then was consensus swinging toward Mellanby's
position? Scottish pediatrician Leonard Findlay speculated: "The

vitamin theory is certainly seductive, and to-day is the most popular one, but it owes its acceptance in great part to its novelty,
though the prevalence for many years of a general idea that rickets
is a dietetic disease has also materially helped."[51] Mellanby countered that novelty and hype worked *against* his findings prevailing.
Most of the skeptics, he suspected, would "be whole-hearted be-
lievers in the importance of accessory food factors in child nutrition were it not that they are blessed with the name 'vitamines.' "[52]

In 1921, Findlay, a "whole-hearted believer" in environmental causes for rickets, surveyed the state of affairs and concluded
there were "2 sets of experimental workers, apparently carrying
out comparable experiments, and yet obtaining very different results." Hope tempered his disappointment. "It is a pity that some
means of bringing them together could not be found so that the
error, for some error there must be, under which one of the groups
is laboring, could be discovered."[53] As if to make Findlay seem a
minor prophet, two research projects underway in Wisconsin and
New York City would square the circle and reveal that both sets
of workers were closing in on one paradoxical truth—the func-
tional link between UV light and the active rickets preventive in
"accessory food factors." In New York, Alfred Hess and Mildred
Weinstock had been studying the effects of UV light in preventing rickets. They found that rats remained rickets free if fed human cadaver skin that had been exposed to UV light. Exposing
other foods, from linseed oil to lettuce, had the same effect. At the
University of Wisconsin, Harry Steenbock and Archie Black had
come to the same insight. Both teams published their research,
weeks apart, in 1924.[54] Steenbock had submitted his paper, but
asked to delay publication until his patent application for the irradiation process was further along. But when a colleague alerted
him that there were two studies in the pipe, he released his study
for publication.[55] Steenbock went on to secure a patent for the
process in 1928, a subject for a later chapter.

Well before Steenbock and Hess explained the relationship between UV light and dietary sources of vitamin D, less strident and
optimistic voices promised the closure Findlay despaired of the two

camps ever finding. In 1922 *Nature* published a brief editorial on a recent essay by Findlay, optimistically noting: "It is fortunately no longer necessary to try to decide which of these two views is correct for, as so often happens, it is now pretty clear that both are right, and, which deserves less notice, that both are wrong. The two propositions are indeed not contradictory but complementary."[56] Still, it is tempting—a temptation to which historians have routinely succumbed—to ignore the deep divisions between advocates for the "environmental" and "dietary" explanations, a division so deep that British writers tended to define it geographically, with Findlay, Noël Paton, and the environmentalists being the "Glasgow School," and Mellanby and others (including those in Wisconsin and Baltimore) the "London School."[57] It is easy for us, knowing the "obvious" answer to the riddle, to be taken aback at the two sides' failure to see it: "both sun and diet."

But the division was meaningful and deep. At a 1922 forum of researchers including Findlay and Hutchison, Mellanby praised both sides for bringing attention to rickets and having a determination to address it, but anticipated that were they asked to propose plans for eradicating rickets, "our recommendations would be so different as to suggest chaos," with some seeing "housing schemes as a panacea," others recommending massage and electric treatments, or "more sunlight and ultra-violet rays." Still others, presumably including Mellanby himself, would argue that "proper feeding of children would settle the problem."[58] Observing this impasse, it is tempting to see in Steenbock's and Hess's discovery the key to ending what the philosopher of science Thomas Kuhn called an incommensurability crisis in scientific theories. Of course, the solution would be as if Copernicus had successfully argued for a cosmos simultaneously heliocentric and geocentric.

If both sides were right, why then has the history of vitamin D so often highlighted Mellanby and the others who favored a nutritional explanation, as a quick dip into online metadata makes clear? At least four factors explain this preference. First, making the project about finding an antirachitic *substance* (as opposed to identifying environmental *conditions*) fits vitamin D more neatly alongside the other vitamins and the story of the rise of nutritional science. Vitamin D is unique among the micronutrients in having two radically different sources; focusing on the nutritional side "normalizes" its story, making it look more like A, B, C, and E. It also makes a better fit with the quick guide to modern biomedical discovery: identify, isolate, inoculate. The search McCollum and Mellanby undertook for a dietary factor preventing rickets "looked like" normal modern biomedical research as conducted by the great microbe hunters in their quest for chemical magic bullets. That is a story of white lab coats, test tubes, and chemical analyses and the search for an identifiable, isolable, and eventually manufacturable substance. The ability to manufacture a reliable source of liquid vitamin D made all but inevitable the public health response to rickets: not sun baths but vitamin D–fortified milk, requiring tons of industrially produced "liquid sunshine." Nutrition won; hence the third, and snarkiest, argument for the subsequent emphasis on "the London School": history, they say, is written by the victors.

The fourth reason turns the lens toward the environmental side. Although Findlay was right that both sides often carried out "comparable experiments," much of his side's project was tinged with slightly antique associations whose negative valence only grew stronger with the ascent of the "golden age" of modern biomedicine. The so-called Glasgow School often decried want of sunshine in causing rickets, but those calls blended into a litany of "hygienic" solutions more in keeping with Victorian hygienists with their vitiated air and miasmas, than with "modern" biomedical research practice. In the modern age, McCollum was Sinclair Lewis's scientist hero Martin Arrowsmith, while Findlay came off more like Arrowsmith's superstitious counterpart, Almus Pickerbaugh.

A quick review of the research supporting the "hygienic" side, however, shows a body of work as scientific as the nutritionists'; their extended laboratory experiments, comparative geographic surveys, and human clinical studies provided important insights and crucial data. In terms of geographic studies, two bookend the most relevant years. Despite its simple methodology, Theobald Palm's 1890 survey of medical missionaries provided convincing evidence that the quantity and quality of sunlight was an important factor in the "Geographic Distribution and Etiology of Rickets"—more important than diet in the case of geographic distribution.[59] Thirty years later, Harry S. Hutchison, another Scottish medical practitioner in Asia, conducted a powerfully instructive survey of rickets in the Nashik district of Mumbai, then the "Nasik" district of the "Bombay Presidency." The poorest children, whose parents were agriculturalists and laborers, had uniformly inadequate diets—high in carbohydrates but "deficient in animal fat and vegetables." While they suffered from many conditions of malnutrition, including scurvy and anemia, they were all but free from rickets. Paradoxically, rickets appeared in over 10 percent the children of "comparatively well-to-do" families, despite a diet rich in proteins and fats. The explanation rested in cultural practices arrayed across gender and class. The young mothers from better-off families were able to practice *purdah*, remaining indoors with their infants, while poorer women continued to work, usually outdoors, and brought their infants to the fields with them.[60] Hutchison made explicit the link between conditions in rural India and industrialized British cities in the West.

As for experimental studies, the environmental camp matched the nutritionists' animal studies in quantity and rigor, sacrificing another hecatomb of rats and puppies on the altar of the sun. Findlay himself was responsible for one of the most important animal studies, an experiment published in 1908 in which he tried, unsuccessfully, to induce rickets in collie puppies by confining them and feeding them an all-cereal diet. His experimental animals "invariably wasted, became marasmic, and died," but never became rachitic. Three of his control animals, who received milk in addition to the cereal diet, did develop rickets, "one of them however, less severely than the other two." This lucky dog was fed and kenneled with the other controls, but because it "was ultimately intended as a companion" it was "exercised by its prospective owner once or twice daily." Suspecting the outdoor activity was responsible for this dog's comparative health, he allowed it "to get still more exercise for about a week, when all appearances of rickets practically disappeared." Findlay did not say who the prospective owner was, but the dog's response to fresh air and exercise reinforced Findlay's conviction that exercise, more than anything dietary or environmental, was the single most important factor in preventing rickets.[61] Nine years later he and his research partners at Glasgow published the results of a subsequent experiment on seventeen collies from two litters. Those kept completely confined developed rickets, while those allowed free exercise in the open air remained healthy.[62]

At the University of Wisconsin, meanwhile, Harry Steenbock and his graduate advisor, Edwin B. Hart, demonstrated that lactating goats kept indoors lost skeletal calcium when fed "old dried roughage," but retained calcium when fed sun-drenched "green pasture"; this was Steenbock's first experiment pointing him toward his discoveries functionally linking UV light and vitamin D, and from there to the patented process that would bring great wealth and acclaim to him and his university.[63] Steenbock's patented process would provide the potent synthetic vitamin D that would be used in dozens of experiments with human subjects, including the "extremely interesting . . . colored infants" in whom Joseph Stokes and his team induced rickets in 1934.

4

Testing the Cure

The Science of Rickets II

The "vitamin hunters" argued for decades over whether light or diet was the more powerful "specific" for rickets. For those conducting therapeutic studies in both community and clinical settings, the divide between the diet and light "camps" was far less sharp. For decades hospitals and sanitaria had incorporated heliotherapy into their general systems of therapeutics, blending dietary and light-based interventions to treat all manner of conditions, including rickets. Auguste Rollier opened Le Chalet, the first of his heliotherapy clinics in the Swiss Alps in 1903, and while the clinic focused on tuberculosis, in 1916 Rollier offered convincing empirical evidence of a sun cure for rickets.[1] German pediatrician Kurt Huldschinsky built on this with a clinical experiment published in 1919 that tested "artificial sunlight" from quartz-mercury lamps. Not only did he confirm that "artificial" exposure to UV light cured rickets, Huldschinsky proved that shining the light exclusively on one arm of a rachitic child produced healing in both arms, evidence that would be crucial to the science of fortifying foods with vitamin D.[2]

Many of these clinical studies demonstrated a certain laxness about constraining variables. In 1920, nutritionist Harriette Chick, from London's Lister Institute, led a team of researchers in a two-year study of hospitalized children in two Viennese clinics.

Chick's goal was to study the effect of two diets in the prevention of rickets, but the study's most significant finding concerned
the sun. The clinics both featured ample natural lighting, and patients routinely received sunbaths whenever the weather permitted. Half of the infants in the study received whole milk from the
country, augmented with cod liver oil; the other infants were fed
milk from Viennese dairies, with no cod liver oil added. None in
the first group contracted rickets, while almost all the others did.
However, their symptoms became pronounced only during winter
and early spring, and by midsummer they were healed.[3]

By 1921, Alfred Fabian Hess had as good a sense of the "two
camps" issue in rickets research as anyone, having navigated both
sides for several years in significant studies, often with Lester
Unger, a colleague at the College of Physicians and Surgeons.
"There are two widely divergent theories," they observed, "the
hygienic and the dietetic, the pendulum swinging sometimes in
the one and sometimes in the other direction." Hess was also well
aware of the methodological fuzziness in many clinical studies—
leaky experimental protocols, such as Huldschinsky muddling his
UV lamp studies by providing dietary supplements, including cod
liver oil.[4] Hess would die a dozen years later, at the age of fifty-
eight. In a remarkable two decades, he published over two hundred papers, most notably laboratory-based studies of rickets and
scurvy.[5] He studied the roles of both diet and light in rickets. He
was equally catholic in his methodology: in addition to conducting studies with laboratory animals, he walked both sides of the
divide separating therapeutic community-based studies and the
more troubling clinical controlled nontherapeutic trials.

Throughout the 1910s, Hess had been studying scurvy among
poor children in New York City, at times withholding orange juice
from test subjects to produce symptoms to cure. In 1917 he published his first study of rickets: a trial of cod liver oil, entitled
"Prophylactic Therapy for Rickets in a Negro Community." He and
Unger wanted to study "a negro community," because they believed "that it can be safely stated that over 90 per cent. of the colored babies have rickets," providing "an excellent opportunity . . .

for judging therapeutic results." The Columbus Hill district, a predominately black neighborhood on Harlem's West Side held a special attraction: the Henry Street Settlement and National League had recently conducted a social survey of the neighborhood, providing data on "economic and living conditions, as well as the morbidity and mortality."[6]

Hess and Unger selected eighty infants for a six-month community-based therapeutic study spanning winter and spring of 1917. Clinical examinations at the start and end of the trial tracked the symptoms of rickets in each child; visiting nurses made frequent follow-up visits in the children's homes to make sure mothers were administering the oil. The study confirmed cod liver oil's effectiveness, preventing or curing rickets in over 90 percent of infants receiving the oil for the full six months, and in over half of those who received oil for only four months. Those who received no oil almost uniformly showed advancing signs of rickets. Hess urged that cod liver oil dispensaries be established in "large cities in the negro and the Italian districts, for rickets is almost as prevalent among the Italians as among the colored people."[7]

Hess and Unger noted there were "compensating advantages" in using African Americans as subjects. "As the negro, in the choice of his home, is limited to certain definite localities, our follow-up work was not hampered by the great difficulties which usually beset those who attempt to keep track of people in this great city." Still, housing segregation alone did not provide enough "compensating advantage" for the kind of scientific inquiry Hess preferred, a "systematic study of a group of infants carried through for a long period." This, he argued, "is impossible in the hospital, in the dispensary or in the home, but can be carried out only in an institution where the hygienic and sociological conditions are alike for the entire group, and where all factors are under absolute control."[8] Of course, Hess had such an institution in mind, the Hebrew Infant's Asylum, where for several years he and Dr. Sidney Haas were attending physicians.

The asylum took care of about four hundred children under the age of five, "who have been committed by the city . . . generally

on account of the death or the destitution of their parents."[9]
These children provided a font of test subjects for Hess and his
colleagues: he developed and tested a catheter for extracting
gastric juices, tested pertussis vaccines, and conducted his most
significant research before he turned his attention to rickets—a
nontherapeutic clinical study of scurvy, frequently a bane of orphanages, in which he and research partner Mildred Fish withheld
orange juice until his subjects developed the bleeding lesions that
signaled the onset of scurvy.[10]

In 1918, within months of completing the Columbus Hill study,
Hess and Unger undertook what would become their second major study of rickets in children, a clinical experiment involving
approximately one hundred infants. In this experiment the researchers crossed the threshold separating therapeutic community studies and controlled experiments with laboratory animals.
As infants arrived at the asylum, they were put on one of four diets: "In all cases there was but one deficiency in the diet, which
was adequate . . . in its caloric content" and would not produce
scurvy (in their notorious scurvy studies, Hess and Fish had been
equally concerned to provide cod liver oil so as to prevent rickets
in the children they were giving scurvy).[11] Sure enough, rickets
developed, even among those children whose diets contained "the
fat soluble vitamin" as vitamin A was still called (McCollum having not yet isolated vitamin D), administered for the most part in
the form of milk. Infants in the study received cod liver oil only if
their rickets became severe and unresponsive to higher ingestion
of the dietaries being tested, as in one of the cases included in the
report: "Case 1: J.C., a baby admitted when 1 week old, without
rickets, was given daily when 3 months old a quart of milk containing fully 3 percent. of fat. By the middle of November when he
was 5 months old, he had developed marked rickets. His general
condition had improved during these months, he having gained
2 pounds in weight. Rickets persisted for ten months, in spite of
this large amount of milk, until cod liver oil was given in October."
This study was only a modest success, inasmuch as it did *not* show
that "the fat soluble vitamin" cured rickets, but that something

in cod liver oil did. More significant, it seems to have convinced Hess that even in his own institution, the factors were not under complete control.

One factor completely out of his control was how the public would perceive his research. A year after Hess and Unger's study appeared in *JAMA* (the nation's flagship medical journal), the American Left's flagship journal, *The Nation*, published "Orphans as Guinea Pigs," a scathing three-page editorial by Konrad Bercovici, decrying Hess's scurvy and rickets studies. "No devotion to science," his summation argued, "no thought of the greater good to the greater number, can for an instant justify the experimenting on helpless infants, children pathetically abandoned by fate and intrusted to the community for their safeguarding."[12] Apparently the rebuke stung; within a month the *New York Times* published a puff piece on the asylum that included an extended glowing appraisal of Hess's rickets study there, a study with "a larger scope than merely furthering the health and wellbeing of the little inmates intrusted to the care of the Home . . . but to the children of the poor who have to dwell in the crowded tenements of our municipalities."[13]

The medical press also responded promptly: *American Medicine*, a relatively new journal founded to be "owned and controlled by the medical profession," agreed that human experimentation should be "limited to volunteers, or to such children as may be utilized with parental consent." But, the editors argued, the state assigns orphanages "responsibility *in loco parentis*," so "parental" consent is not an issue. Nor do Hess's studies rise to the level of vivisection, which they defined as "experimentation by means of drugs and foreign material introduced into the human body." The experiments at the Hebrew Infant Asylum were more like the "variation in feeding as is today practiced in many households" where parents are ignorant of the causes of rickets or scurvy. It was crucial to put scurvy and rickets "safely in the category of preventable diseases"; thus "the experiments of Hess and others are not to be regarded as indefensible."[14] Whether Hess was convinced is not clear, but for whatever reasons, most of his subsequent experimental vitamin D research—by far his most significant studies—involved rats, not children[15]

Where Hess saw himself as a practicing physician and scientist, Martha May Eliot seemed to live up to her youthful hope of being "some kind of social doctor." Accordingly, her famous studies of rickets, conducted in the same years that Hess was busy inducing rickets in rats, sought to test "whether rickets could be prevented in a community by the intensive use of cod liver oil and sunlight." Trained in medicine at Johns Hopkins, Eliot served in Yale's pediatrics department from 1921 to 1935, a position she held even after joining the federal Children's Bureau in 1924 to direct its child hygiene division. For years she juggled the New Haven–Washington, DC, split, spending one week a month in the capital but maintaining the home in New Haven she shared with her life partner, noted pediatrics researcher and reformer Ethel Collins Dunham. Eliot eventually rose to chief of the Children's Bureau, then left in 1956 to direct the child and maternal health department at the Harvard School of Public Health—perhaps a soul-satisfying move, since Harvard's Medical School had denied her application in 1914 on account of her gender.[16]

Eliot's study of rickets in New Haven began in the fall of 1923, with the cooperation of the Federal Children's Bureau, the Yale School of Medicine, and New Haven health organizations. For the next three years, she and her staff of visiting nurses, physicians, x-ray technicians, and social investigators identified and sought to enroll every child born in a district populated by over thirteen thousand, "one third of which were negroes, and two thirds a mixed population composed of Italians, Irish, Polish and Americans." Participating mothers brought their children to the office once a month for examinations, including x-rays. Her first interim report, published in *JAMA* in 1925, assured nervous AMA members, "No new baby is brought to be examined without the knowledge and consent of the family physician.[17] Mothers received training in how to administer the oil: "With her left hand she opens the baby's mouth by pressing the cheeks together between her thumb and fingers. The oil may then be poured little by little into the baby's mouth. If the mouth is not held open until the oil entirely disappears, the baby will spit out what is left." They were also given detailed instructions regarding sunbaths for their infants. Of the 216 infants who remained in the study's first year, 186, or 86 percent showed signs of rickets by clinical exam, with even higher rates as diagnosed by x-ray; but comparisons with three control groups gave solid proof that the oil and sunlight reduced the severity of the rickets.[18]

The study suggested "a slight degree of early rickets is well nigh universal in our climate and in our state of society." This incidence, even among the group treated with oil and sunlight, raised a question she admitted she was "not yet prepared to answer," whether perhaps "this slight degree of rickets must . . . be considered normal."[19] To answer this, Eliot conducted a similar study in Puerto Rico, which concluded what observers had known all along, that rickets was rare in the tropics, except among those kept out of the sunlight. Of the nearly six hundred children given both clinical and x-ray exams, fifty received a positive diagnosis, though forty-six were designated as "slight degree" and only one "of marked degree"—a six-month-old boy who had been born in a stone cellar "and was never taken out of the cellar for fear he would 'catch cold.'"[20]

Despite the continued flurry of community surveys, lab experiments, and clinical trials—and the growing mountain of data they produced—the project to reduce and cure rickets was at a bit of an impasse in the mid-1920s. Antirickets campaigns had but one effective tool (two, if you include the sun): cod liver oil had been employed against rickets for a century, despite science having no adequate explanation for the cause of its effect. Insights into how cod liver oil worked did not change its nature; knowing that sunlight prevented rickets didn't brighten the cities. Researchers could seek the best way of administering the foul-tasting concoction; they could rail impotently against urban conditions that held children in sunless tenements. But if the actual prevalence of rickets in northern cities was even close to the 80 to 90 percent reported in study after study, a new intervention would be required. Effective rickets intervention was not going to take place across the physician's desk, one patient at a time. It would involve population-wide interventions of one sort or another. The community studies highlighted some of the roadblocks: providing free cod liver oil for poor families, maintaining education campaigns and brick-and-mortar clinics—these relied on ongoing infrastructural and professional investment. And the problems with cod liver oil itself were significant: in addition to its taste, its potency and shelf life were unpredictable, as was, recent experience in the Great War suggested, its availability. What was needed was a reliable means of putting predictable doses of vitamin D in the diets of all children.

The Steenbock process for irradiating food with ultraviolet light offered those reliable means. First patented in 1928 by Harry Steenbock of the University of Wisconsin, the process promised cheap and palatable vitamin D fortification. The cutting edge in rickets research quickly moved on from quaint old cod liver oil to Viosterol and the best way to harness the power of UV radiation to turn regular foods into powerful rickets preventives.[21] So many substances, from grains to oil to milk, could effectively deliver manufactured vitamin D: Which was the best for rickets prevention in the United States? In the next chapter, I take up that often-wild quest, the array of foods and methods researchers and industrial interests considered on their way to the *seemingly* inevitable choice of milk, largely settled by the early 1930s. The promise of universal vitamin D fortification using the Steenbock process spurred both animal and human studies to answer anew the questions that Hess and Eliot had sought to answer about cod liver oil: How was the potency of irradiated foods to be assessed? What was the effective dosage for prevention? For cure? "Obviously," one researcher concluded, "in the present state of our knowledge the only true evaluation of antirachitic potency for infants of a new vitamin D fortified milk must be actual clinical trial."[22]

Many of these studies were undertaken with funding from the Wisconsin Alumni Research Fund, or WARF, initially established by Steenbock and other researchers at the University of Wisconsin to manage the proceeds from patents arising from their studies. The food industries saw immediately the prospects of the Steenbock process. Even before Steenbock received his patent, Quaker Oats secured exclusive rights from WARF for use in cereals. In 1928 WARF granted contracts to four major pharmaceutical companies, guaranteeing lucrative royalties and quality control. The dairy industry was especially interested early on, and WARF, located in the heart of "America's Dairyland" supported research into several ways of fortifying dairy products.[23]

By the early 1930s, evaporated milk had come a long way from its origins as a reliable source of nutrients for soldiers and sailors. In the quarter century after Carnation started promoting its product as "the milk from contented cows," US consumption of the canned milk tripled.[24] Canned milk's fortunes were also buoyed by the rising popularity of infant formula, so the evaporated milk people were keenly interested in the possibility of fortifying their product with the powerful antirickets vitamin.[25] WARF licensed the Steenbock process to five evaporated milk manufacturers, who, through their quickly established Irradiated Evaporated Milk Institute, coordinated with WARF and researchers to study and no doubt provide scientific validation of the value of a fortified evaporated milk.[26]

It was to WARF that researchers at the University of Pennsylvania Medical School and the Children's Hospital of Philadelphia turned in 1933 to fund their proposed trial of irradiated evaporated milk. The previous year, the Wisconsin fund had supported a study by CHOP researchers testing the antirickets power of various irradiated milk products "in parallel series groups of human infants and rats."[27] The human infants were orphans at St. Vincent's Hospital. Dr. Dorothy Whipple proposed a more controlled study with twelve infants—six receiving vitamin D milk, six receiving no antirachitics. WARF officials wanted a more statistically robust study and encouraged CHOP to double the number of babies in the study. J. Claxton Gittings, physician in chief at CHOP, agreed, proposing that a few more than twenty-four be enrolled: "In dealing with

babies," he observed, "we must be prepared to withdraw several of
them in the event of the intercurrent disease, the sudden decision
of parents to remove them, etc."[28]

::

On January 2, 1934, Philadelphia's weather was clear and season-
ably cool, with temperatures reaching a high of thirty-nine degrees.
We may assume, then, that dozens of patients at the Children's
Hospital of Philadelphia spent part of their day in the open air of a
rooftop sun terrace.[29] But in a ward of the newly opened Benjamin
Rush Building that day, Dr. Stokes's "interesting group" of infants were not rolled out into the sun. Instead, the twenty-four
African American boys—boys, to simplify urine collection—were
weighed, examined, and given the first of five monthly x-rays and
blood draws (for phosphorus and calcium assays). Next, each subject was placed in his own glass cubicle in a well-ventilated but
sunless ward of the hospital, where he would remain for nineteen
weeks of darkness and deprivation.[30] Upon admission, three of
the infants showed very mild signs of rickets. For the first month,
all the babies were kept out of any sunlight and received "non-
irradiated" milk "to eliminate the effect of any previous antirachitic
therapy." After the control period, thirteen exhibited clear x-ray
evidence of rickets. Any child diagnosed with rickets after the control period was put on the experimental treatment: vitamin D–
fortified evaporated milk, provided free of charge by the member
manufacturers of the Irradiated Evaporated Milk Institute. The
other half continued to receive unfortified milk. Although some
developed marked signs of rickets, none were dosed with cod liver
oil or concentrated vitamin D drops, though Hess and Eliot had
demonstrated cod liver oil's curative efficacy. By the end of the
study, x-rays showed fourteen of the subjects still exhibiting active
rickets, though some showed improvement.[31] The study's authors
concluded that the evaporated milk in the study "appeared to be
an adequate agent for the prevention of rickets in negro infants."

Adequate for *preventing* rickets, but, they continued, "it also appeared to be an unreliable agent for the cure of rickets in negro infants."[32]

The evaporated milk people were apparently pleased with the results of this and other studies. That spring the AMA's Committee on Foods had approved the use of irradiated milk as an antirachitic, a fact that Carnation Milk announced on the Carnation "Contented Hour" radio program in June, in an episode featuring the "world famous coloratura soprano, Madame Galli-Curci."[33] That same month Carnation contacted WARF to fund another study at the Children's Hospital. Dr. Harry L. Russell, director of WARF, wrote to Dr. Joseph Stokes at CHOP in June 1934 that the evaporated milk people "were of the opinion that it would be highly desirable to have some additional work undertaken with *white* children to see what the susceptibility of the dominant race is when compared with the colored race."[34]

Difficulties ensued in replicating the 1934 inpatient study. The hospital ward previously used was not available, and researchers struggled to enroll babies. The desired race of the subjects for the new study may have played a part (the Irradiated Evaporated Milk Institute had gone so far as to specify "that all babies used are to be white babies even omitting any that are Italian").[35] The team managed to secure nineteen white infants for a four-month study "to determine the prophylactic antirachitic value" of irradiated evaporated and whole fluid milk. But this was a very different study: unlike the previous experiment with Black babies, this time there was no attempt to induce rickets in the subjects. The goal was to study irradiated milk's preventive power, not its ability to cure. Although fifteen of the infants were kept in a sunless ward and not taken out for fresh air, they received antirachitic milk every day. In the end the study found that both milks "appeared to be equally efficacious in preventing the development of rickets."[36] The study morphed from the original four-month inpatient study of a handful of white babies into a multiethnic outpatient study of about one hundred infants. After the first year, the study grew to over two hundred infants, who were followed for up to four years. The scope of the study was also expanded to track growth and general health.[37]

The 1934 study at Children's Hospital took place ten years after the *Nation* exposé of Hess's experiments at the Hebrew Infant's Asylum and in the wake of decades of activism against animal vivisection. But this attention had not produced professional standards or formal guidelines to curtail nontherapeutic experimentation on children. The experiment may mark the nadir of rickets clinical trials, but it was far from unique; it was no anomaly in a more enlightened epoch. The team of researchers there undertook at least two nontherapeutic trials before shifting to outpatient parallel preventive and curative studies. Neither their published reports nor internal correspondence suggests that this shift was due to anything but the difficulty of securing "clinical material" for inpatient trials. A report from the AMA Committee on Foods, drafted around 1935, summarized about a dozen US and Canadian clinical and outpatient studies of Vitamin D milk products from the late 1920s to the mid-1930s. It was fairly common in these studies to begin the trial with a control period of a month or so, during which the subjects were given no antirachitics; uniformly, this resulted in marked increase in rickets diagnoses among the subjects, providing adequate experimental material for the trial.[38]

One study that did *not* show up in this meta-analysis had been conducted in New Orleans by Dr. Rena Crawford in the clinics of that city's Child Welfare Association. For over three years, Crawford's team tracked the development of rickets in the first year of life among hundreds of the city's poorest children, whose mothers were provided with cod liver oil and instructed in its use. Crawford's study failed to meet the criteria of sophisticated experimental studies, partly for lack of controls: "We wished to put half our babies on cod liver oil at ten days and to give the other half no oil till they showed symptoms of rickets," the study's authors reported, "but the management of the C. W. A. would not allow us to do that, but required that all babies delivered by the Association should receive the oil at ten days old."[39] Still, the study confirmed some important points: that measuring dietary compliance in an outpatient setting involved more faith than certainty; but to the degree that dosage *could* be accounted for, cod liver oil was an effective preventive. The incidence of rickets the study found—in one of the nation's most sun-drenched cities—should have been a sobering corrective to the growing myth of rickets as a problem of dark-skinned children living in northern climes to which they were, according to the science of the day, evolutionarily ill suited. Nearly two-thirds of the infants Crawford's team followed developed x-ray evidence of rickets, with those receiving the most cod liver oil, and on something near a regular basis, faring significantly better (just under half receiving a positive diagnosis). This southern study is instructive in a number of ways, not least of all its message that even if its findings were weak, useful, patient-centered research might best be undertaken away from the imperative of corporate and institutional priorities.

The New England Journal of Medicine published a literature review of these studies in 1938, which spelled out those guiding priorities. Earlier studies, between 1934 and 1937, were inpatient studies of rachitic children. The later studies "have been predominantly out-patient investigations of normal infants." The change, they assumed, was "due partly to the unavailability of suitable rachitic infants." More significant, they concluded was "a change in point of view." Earlier studies prioritized testing the fortified milk's curative power in "an effort to put vitamin D milk to as severe tests as possible." Having accomplished that, later studies measured "the power of vitamin D milk to prevent the appearance of rickets in infants living under home conditions."[40]

Most of the studies of the 1930s did not test that nasty old preventive, cod liver oil, but the new miracle of fortified milk—the presumed perfect delivery device for vitamin D, as proven by dairy industry–funded research. The ethical failures demonstrated in the CHOP studies put them squarely in the bleak company of Tuskegee and Willowbrook. WARF-funded clinical trials inflicted real harm—and in a quixotic search for fixed truths where none lay: optimal doses of an artificial nutritional adjuvant when the underlying metabolic factors could never be controlled for; universal diagnostic criteria where individual differences in age, rate of growth, bone density, prenatal factors, and a host of other confounders precluded such specificity. Viewing example after example of false precision in these researchers' tables, and reading their sanguine conclusions, one must question to what degree these researchers believed their own spin.[41]

: :

This survey of the science of rickets has taken us up to rooftop solaria, where each sunny day, children were naturally infused with vitamin D, and down into the specially fitted sunless ward where the infants selected for a clinical experiment were dosed with evaporated milk. The tour concludes with a visit to the orthopedic surgical ward, where those children whose legs had been permanently distorted by rickets underwent surgery or weeks of immobilization while various orthopedic devices or plaster casts reshaped their legs.[42] This chapter has traced the central role rickets played in the rise of modern nutritional science. It concludes with a brief nod to the central role rickets played in the rise of pediatric orthopedics.

Rickets could leave its stamp on a child's skeleton in a number of ways, including curvature of the spine, "pigeon breast," pelvic distortion, or curved lower leg bones. But well over half of rachitic cases orthopedists treated bore specific deformities of the long bones: bowlegs or knock knees ("*genu varum*" or "*genu valgum*," respectively).[43] Healing these bent limbs was beyond the power of cod liver oil or viosterol. The particular approach to straightening a child's rachitic limbs depended on many variables, age and degree of curvature chief among them. Nonsurgical interventions included orthopedic shoes to shift the child's stance or the more powerful pressure of braces and plaster encasements; all these guided growing bones toward correction. Surgical interventions were of three essential types. The oldest is osteoclasis, the breaking of the curved bone without opening the leg surgically, an operation that prior to the advent of antiseptic surgery in the 1870s would pose grave risks of fatal infections. Osteoclasis was safer regarding infection but far from precise in terms of where the break would occur. Gaining direct access to the bone through even a small incision permitted osteotomy: focused breaks, or— more frequently—partial breaks, or the removal of wedges of bone to facilitate reshaping.[44] Surgeons continue to refine surgical interventions for rachitic legs, but today, most still rely on some form of osteotomy.

For much of the twentieth century, the most common procedure for surgical correction was the subcutaneous osteotomy, an operation developed by William Macewen at the University of Glasgow. Macewen studied with Joseph Lister in the years Lister was developing his system of antiseptic surgery. In 1875 Macewen performed his first antiseptic osteotomy and over the next ten years performed over 1,800. His operation advanced the state of the art considerably. Previously, surgeons usually cut or broke the bone completely, using chisels and saws, requiring large incisions and often producing significant bone and muscular trauma. Macewen invented a special chisel he called an osteotome—a slender tool, with its blade beveled on both sides. His operation, always under anesthesia, started with a small incision through just the skin, running parallel to the muscle tissues beneath. He then inserted the osteotome into the incision and parted the muscles to push the tool through to the bone. Twisting the blade ninety degrees placed it in position on the bone to make whatever cut the particular case required—a partial break or the removal of a wedge. The wound was then closed, and the leg put in a splint to reshape the bone.[45] In 1880 when he published a textbook on the procedure, he had "operated on 557 limbs, belonging to 330 patients." In many cases he performed more than one osteotomy on a given limb with complex distortions. Clearly, although many of his patients were adults or adolescents, in most cases they acquired their condition from childhood rickets. Macewen was explicit in attributing his patients' condition to rickets, even though

many developed the condition in adolescence—so-called late rickets or osteomalacia. Two-thirds of his patients "were affected with
genu valgum," and many of the remainder were bowlegged or had
other rachitic deformities.[46]

Macewen did not publish detailed statistics on surgical outcomes, though he reported that in all but eight cases, "the wounds
healed by organization of blood-clot without pus-production."
One patient developed gangrene—"a bow-legged patient," who on
his second night "rose from bed and fell," twisting his splint and
bandages. Because the patient "foolishly concealed the accident,
though suffering great pain," nurses did not discover the injury
until ten hours later. His gangrenous leg was amputated. Macewen
reported that only three of his 330 patients died, though not, he
claimed, as a direct result of their operations. He argued that
the "constitutional" shortcomings of those patients were largely
to blame. "The majority were in a low state of health," with tuberculosis rampant and bronchitis common. This was consistent
with the general health of the poor Glaswegians he treated at the
Glasgow Royal Infirmary and abetted by his determination not to
"filter" his patient pool. Though he initially selected patients in
good health, "it soon became apparent if all those who were weak
and feeble were to be rejected, the operation would be denied to
those in whom it would be of greatest benefit."[47]

Long before the osteotome was the splint. The written history
of nonsurgical techniques for correcting rickety deformities goes
back to the second century CE and the Greek physician Soranus
of Ephesus, who approved of the ancient practice of swaddling
infants in order to help babies to "grow straight." He suggested
swaddling infants, including the upper body and arms, for the first
few months, a practice modern practitioners might argue could
actually cause vitamin D deficiency.[48] Francis Glisson's 1650 treatise, for which generations of historians inaccurately credited him
with "discovering" rickets, contains detailed descriptions of the
orthopedic appliances he developed in his practice. These included
splints using iron plates, whalebone, and woolen padding. He de-
signed a splint with two iron plates that could rotate around an

axis, allowing the wearer to bend their knees. He also developed the "Glisson sling," a method for suspending a swaddled baby from the shoulders and torso, leaving the affected lower body and legs free to dangle. In order that "the parts may the more be stretched," Glisson recommended—at least for "the more robust children"—that physicians hang "leaden shoos upon the feet, and fasten weights to the body, that the parts may be the more easily extended."[49] The word "orthopedics" derives from French physician Nicolas Andry's mid-eighteenth-century treatise on "the Art of Correcting and Preventing Deformities in Children." Andry emphasized splinting rachitic bones, and in fact his book was illustrated with the now-iconic image, found in the heraldry of countless orthopedic organizations and practitioners' logos, of a crooked tree bound to a straight post.[50]

If rickets played a central part in the rise of modern orthopedics, so too did the developments in our understanding of vitamin D—the science of rickets—shape the mission and practices of orthopedic institutions for children. From the mid-nineteenth century, when urban health experts noted what Buffalo physician Irving Snow would later decry as "a constitutional deterioration from a great and sudden change in environment," hundreds of institutions, some charitable, others municipal or state run, arose to treat and rehabilitate children with all manner of orthopedic conditions.[51] Children with the aftereffects of rickets always made up a significant portion of their patient population, so their mission, their methods, and their very existence were tied to the history of rickets before and since the era of vitamin D fortification, before and since the advent of safe and effective surgical corrections, as demonstrated by the history of America's first orthopedic hospital.

In the midst of the Civil War, Dr. James Knight established the Hospital for the Ruptured and Crippled, a twenty-eight-bed facility in his Manhattan brownstone, complete with a conservatory for fashioning orthopedic devices. Knight, a Maryland native, had come to New York in 1835 and established himself as an orthopedist specializing in "surgico-mechanics," or bracing, and as an advocate for New York's poor. Knight advocated nonsurgical treatments, preferring what he called "expectant treatment" instead of surgery's "adventurous" approach. He prescribed fresh air and sunshine; daily life for his patients included "light gymnastics and calisthenics, and instruction, both religious and secular."[52] The emphasis on fresh air dictated both the site and design of the hospital's second building, opened in 1870, "in the country" of what is now midtown Manhattan. The new hospital was a three-story building with beds for two hundred patients, a garden roof "open in the summer and enclosed in the winter," and the entire third floor taken up with a light-filled gymnasium, playroom, and garden. Over the years, however, and against Knight's predispositions, surgery played a larger role in the hospital's operations. A year after moving to the new building, Knight appointed a skilled young surgeon to a prominent position on his staff. Virgil Gibney had trained with Lewis Sayre, one of America's top orthopedic surgeons and something of Knight's antithesis.[53] Surgery rose, solariums faded, and by 1900 the hospital's medical staff of fifty-one included eighteen surgeons. By 1940 the hospital was renamed the Hospital for Special Surgery.[54]

: :

All the hospital wards where rickety children ended up—in the general pediatric ward with its regular trips to the hospital's rooftop solarium, the orthopedics ward where legs bowed by severe rickets were surgically corrected, or darkened wards where children were cruelly tested—all those wards would see their rickets caseloads drop precipitously over the next decades, in large part the result of the campaigns to prevent rickets by providing all children with vitamin D, first in the form of cod liver oil, and then powerful synthetic vitamin D added to most milk American children consumed.

Cod Liver Oil

From Folk Remedy to Proven "Specific"

In 1949, American medical journals ran a striking ad for Mead's Oleum Percomorphum, a fish-oil–based supplement from the infant formula giant Mead, Johnson & Company. A smiling, healthy, bright-white, blond baby crawls energetically across a gray ground, unaware of the ghostly black shadow floating above it—the "shadow of rickets." The scrawny bowlegged specter with a deformed head falls over the baby's legs. "Rickets may be found in apparently healthy and well nourished infants," warns the ad copy, but Mead's Oleum Percomorphum, represented by a school of stylized cartoon fish, "removes the shadow of rickets."[1]

The ad appeared in medical journals across the United States, including the *Journal of the Florida Medical Association*, whose readers in "The Sunshine State," one would think, were not in the market for a dietary supplement to the sun. Although Mead did not advertise their vitamin D nostrum to the general public, other companies at the time did, continuing a long tradition of pitching cod liver oil and high-potency vitamin D products in magazines and newspapers.[2] By the 1930s the relationships between vitamin D, sunlight, and rickets stood on a firm scientific foundation. The old question of whether rickets arose from inadequate light or improper diet had been settled with a firm "both," leaving open

Supplements the sun...
removes the shadow of RICKETS

Rickets may be found in apparently healthy and well nourished infants
due to an insufficient intake of vitamin D plus inadequate exposure to ultraviolet rays.
It is now generally accepted that a vitamin D supplement should be given regularly
not only to infants but to older children and adolescents. Mead's Oleum Percomorphum
With Other Fish Liver Oils and Viosterol is useful for this purpose.

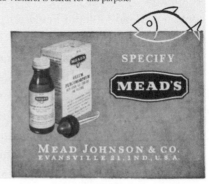
Figure 5.1 Pharmaceutical companies wasted no time exploiting the modern science that explained the power of vitamin D and provided new ways of deliv-ering it. The vitamin D in Mead Johnson's Oleum Percomorphum was produced by the Steenbock process patented by the Wisconsin Alumni Association Research Foundation. *Journal of the Florida Medical Association* 36 (1949–50), back cover.

the markets for products and services to address either cause. As a result, concerned parents had many options for giving their children extra vitamin D. They could choose from an array of cod liver products: oils straight, flavored, or encapsulated; extracts in pill or powder form. Or they could take the environmental approach and increase their children's UV exposure, sending them out to play in the sun or having them bask beneath UV lamps in their homes.[3] The popularity of both dietary and environmental interventions suggests that parents believed preventing rickets was worth the effort, that the shadow of rickets hovering over their children must be kept at bay.

Mead's Oleum Percomorphum, it bears repeating, was *not* cod liver oil, the potent antirickets agent at the center of so much early twentieth-century research. A decade earlier, Germany had invaded Norway, cutting America off from its biggest supply of cod liver oil. No matter, the ad implies: Mead's was better than mere fish oil, since its vitamin D content was enhanced (if not provided entirely) by Viosterol, the brand name for an industrially synthesized vitamin D.[4] Mead's aggressive marketing suggests the market for vitamin D products was expanding even though American dairies had started down the path toward universal fortification, ostensibly making further supplements unnecessary. Previous chapters explored how researchers came to understand cod liver oil's power as a "specific" for rickets. The next chapter focuses on how doctors, institutions, and corporations promoted the raft of products modern vitamin D science created and justified. And since both the science and the marketing started with cod liver oil, the rising and falling fates of that quaint, oft-maligned oil will provide much of this and the next chapter's framework.[5]

: :

The three most important species of fish in the genus *Gadus* are Atlantic, Pacific, and Greenland cod.[6] Historically, the cod's habitat extended over much of the continental shelf of the northern Atlantic and Pacific Oceans. Humans along the edges of *Gadus*'s range quickly came to rely on the bountiful fish. Archaeologists have found the calcified bones of the fish's inner ear—otoliths—in the kitchen dumps of ancient Scandinavians and Native Americans, testament to cod's place in their diets.[7] From medieval times in Europe, dried or salted codfish was a readily traded commodity for markets far from the shore, and the quest for newer and larger fishing grounds fueled much early trans-Atlantic exploration, while salted cod fed the seafarers on those voyages. Cod fisheries provided the livelihoods for fishing towns on both sides of the Atlantic and were an important driver in the Atlantic system of trade, not least for their value to Caribbean slave labor camps. Vast amounts of dried codfish from New England and the Canadian Maritimes provided protein to enslaved sugar workers.[8] The market for the mild-flavored fish continued to rise through the twentieth century until the early 1990s, when overfishing resulted in a near-total collapse of the Atlantic cod population, devastating fishing communities in an arc from Maine to France and forcing a sea change in the fish of choice for both McDonald's Filet-O-Fish and the Burger King Whaler.[9]

A typical six-pound cod, hand trimmed by a skilled filleter, will yield around three pounds of flesh. But before it can be filleted, it has to be gutted. Cleaning that fish yields, among other things, a one-pound head, about a half pound of guts, and—weighing in at around a third of a pound—the fish's largest internal organ, the long, two-lobed liver.[10] Wherever coastal communities owed their existence to the bountiful codfish, as they did from the time of the Norse peoples, they found many uses for cod liver and its oil: in foods, in manufacture, and as a medicine. Extracting the oil was straightforward: English physician Thomas Percival described the process in 1783 as "heaping together the livers of the fish, from which, by a gentle putrefaction, the oil flows very plentifully." Decades later, cod liver oil cheerleader L. J. De Jongh provided a more detailed description. "As soon as the fish is brought to land, the livers are cut out and thrown into vessels, where they ultimately remain." Traditional processing resulted in three grades of oil: pale, or light; light brown; and brown. If the livers were barreled fresh, "the clear oil which comes off . . . is skimmed, poured off, and put into barrels. This is the pale oil." After this, the livers were boiled, "and thence is produced the brown oil." As for "light-brown" oil? It was, De Jongh described, "impaired pale oil," either the product of spoiled livers, or old pale oil that had been stored long past its shelf life.[11]

Neither Percival's nor De Jongh's description fully evokes the olfactory assault of traditional cod liver oil manufacture. Almost any extended discussion, whether in medical treatises or popular magazines, eventually turns to the reeking barrels of its manufactories or the taste of the finished product—or both, as in this 1861 report from a tourist to Patty Harbor, in Newfoundland: "Near by a fish house, there is ordinarily seen a row of hogsheads open to the sun," giving off "smells that none but a fisherman can abide." Closer examination reveals "these casks to be filled with cod livers in a state of fermentation. After a few days in the sun, these corpulent and sweaty vessels yield a rancid, nauseous fluid, of a nut brown hue . . . which I imagine, must have a flavor not unlike that which the invalid finds lurking in those genteel flasks on the apothecary's shelves."[12]

Those foul-smelling casks yielded a product of great utility far beyond the druggist's needs. Cod liver oil's properties have made it an indispensable by-product. Leather processors, from medieval curriers to modern shoemakers, have been major consumers of the most unpalatable "dark-brown" oil, used to soften and finish leather. In the 1850s, the shoe factories of Beverly, Massachusetts, consumed much of the thousand barrels of oil the town's fishermen produced each year. Other longstanding industrial applications include printing ink, various lubricants, lamp oil, animal feed, cosmetics, and paints. Cod liver oil provided the base for the red-ochre paints covering many buildings in fishing villages of Newfoundland into the twentieth century.[13]

Cod liver oil's history as a medication begins close to the docks, with the first people who relied on *Gadus*. Fisher folk swallowed the oil, rubbed it on their aching joints, and applied it to cuts and burns. Centuries later, when medical men had been convinced of the oil's medicinal value, they described cod liver oil as a traditional elixir "from time immemorial."[14] As for its flavor, urbane physicians like Thomas Percival might shrink from the taste, but, he argued, that was immaterial to the traditional consumers, the "Laplanders, and other northern nations," "for habit soon reconciles to the taste of the disgusting viands."[15]

At first, the medical profession was ambivalent about this ancient folk remedy. "There have been few articles introduced into the material medica with more apparent reluctance," Bangor physician Daniel M'Ruer observed in 1849.[16] Still, while popular culture may have imputed to it all the powers of the goddess Panacea, cod liver oil was promising to be sure, and many doctors shared the outlook of one early adopter who argued that folk remedies, "however inexplicable they may be by our theories . . . deserve attention and trial."[17] In region after region, clear evidence of effectiveness eventually overwhelmed professionals' reluctance.[18] Indeed, M'Ruer predicted that since doctors now had given cod liver oil "a place in the catalogue of remedies, it will be as slowly discarded as it has been adopted."[19]

Perhaps this ambivalence contributed to the somewhat foggy history of cod liver oil's acceptance into the Western pharmacopeia. That history has generally addressed three questions, one being medicine's debt to folk practices. The second question involves traditional medical history's perennial search for origins and firsts: which elite male doctor first mentioned cod liver oil? Since *Gadus*'s range does not extend to the Mediterranean, Hippocrates and Pliny could not have tested cod liver oil. Still, both described using dolphin liver, and since modern writers strained for a link to the "Father of Medicine," dolphin would have to do.[20] Candidates for first adopters in modern times are more plentiful, if anecdotal. One tenuous origin story comes from an early twentieth-century physician who claimed his grandfather had used it against consumption; no big claim, but that grandfather, the physician boasted, had it from *his* grandfather that back in the mid-seventeenth century he had given the oil "to some of Prince Charlie's sick soldiers who were entrusted to his care . . . in the '45."[21]

As for the third question—Who first published a clinical report of treating disease with cod liver oil?—conventional wisdom agrees with *Scientific American's* simple assertion from 1871: "It was first introduced into medicine by Dr. Percival."[22] Percival's greatest fame lies in the field of medical ethics. His 1803 *Medical Ethics* outlined a consistent set of guidelines that quickly became the template for codes guiding medical institutions and organizations.[23] Percival developed his code in part as a playbook for the Manchester Infirmary, where he learned firsthand of the therapeutic value of cod liver oil.

Though most of Thomas Percival's medical practice was treating Manchester's merchant families, from his arrival in 1767, he was active in public health, addressing the conditions of the city's poor and seeing to funding and operations at Manchester's large infirmary.[24] In 1782, he published a short essay on "the medicinal uses of the *Oleum Jecoris Aselli*, or Cod Liver Oil," which he extended to treating "rheumatisms, sciaticas of long standing, and . . . cases of premature decrepitude." The infirmary had used the oil for years, "so largely . . . that near a hogshead of it" (roughly sixty gallons) "is annually consumed." Almost half of Percival's report was written by house surgeon and apothecary Robert Darbey.[25] Darbey confined his comments to rheumatism, asserting that some patients, "who have been cripples for many years, and not able to move from their seats, have, after a few weeks use of it, been able to go with the assistance of a stick." Twentieth-century writers have speculated that much of this "rheumatism" was in fact severe osteomalacia, sometimes called "adult rickets."[26] Although he does not specifically mention rickets, Darbey describes curing two children under the age of ten, whose "lower limbs seemed to be a burthen to them, and they had such an appearance of distortion that no hopes of relief could be well entertained." In line with Percival's most famous work—his *Medical Ethics*—and in contrast with the clearly unethical experiments in the previous chapter, Darbey noted his treatments of children were "in compliance with the particular request of their parents."[27]

Darbey was certain the oil would take its place among his age's most indispensable cures. "Except bark, opium, and mercury," Darbey predicted, "I believe no one medicine in the materia medica is likely to be of greater service." And, mindful of the realities of running a charitable institution, he added, "There is one other circumstance which recommends this medicine, which is, that it is a cheap one."[28] If cod liver oil did not fulfill Darbey's prediction about its place in the modern doctor's arsenal, still it remained an important option for treating some of the most pressing conditions of the time, listed by a British doctor in midcentury: "phthisis, chronic rheumatism, rickets, and . . . all cases characterized by deficient action of the system."[29] More significant than rickets' third-place standing on that list is the oil's status as a cure for what might be called "constitutional" diseases. "Deficient action" is a far cry from the later concept of deficiency in a specific nutrient.

Testimonials to cod liver oil appeared in European and American medical journals for much of the nineteenth century, and occasionally in the popular press. The *Providence Patriot* reported in 1825 the case of a patient constantly suffering "most excruciating pains of the back and limbs." Her doctor prescribed a tablespoon of the oil four times a day. After four bottles "the patient had entirely recovered the perfect use of her limbs." The doctor-patient interactions around cod liver oil encouraged the occasional joke: the *New Orleans Times* printed the story of a local doctor who prescribed cod liver oil for his patient, scribbling "Ole. Jec. Ass." (for *Oleum Jecoris Aselli*) to the druggist. The patient recovered fully, forever crediting "that beautiful medicine, the Oil of Jackass, that brought me on my feet again."[30]

The adoption of cod liver oil for rickets followed the familiar pattern of doctors noticing long-standing folk healing practices, testing it in their practice, and being surprised with its potency. An early report, from 1779, related a folk remedy for rickets from the western isles of Scotland: rubbing fish oil (from skate, not codfish

specifically) on the afflicted person's wrists and ankles at night.[31]
In another instance from the 1840s, a physician in Tours was
treating "the rachitic child of a rich Dutch merchant" with iodine,
but to no avail. Then the merchant told him that when his family
lived in Holland, where cod liver oil "was a popular remedy," his
older children had been cured of *their* rickets with it.[32] The doctor
adopted the practice and recommended it to his peers. Similarly,
Edinburgh physician John Hughes Bennett became a convert to
cod liver oil during an extended medical tour of Germany, where
the reports of "complete cure . . . by the oil" were so common as to
"no longer excite attention."[33]

By the mid-nineteenth century, medical excitement about
cod liver oil was reaching its peak, at least in terms of new clinical research. One dedicated promoter, Dutch physician Ludovicus
Josephus De Jongh, crowed that "cod liver oil will supersede every
other means of cure" for rickets "and will accomplish whatever can
be expected or hoped for from any medicine."[34] Some did sound
alarms about misplaced enthusiasm. "There are fashionable medicines as well as fashionable garments," wrote an American physician in 1857. Cod liver oil had been until recently "the great quack
medicine . . . Every doctor who failed to prescribe it was behind
the age." But now, "a few years have passed, and all this delusion is
dispelled." Not that it was harmful, but it would not cure diseases.
"It is simply good nourishment and nothing more."[35]

Why did mid-nineteenth-century medical researchers lose interest in cod liver oil? Perhaps it was the impossibility of defining just what they were testing: the purity, freshness, and hence
the potency of cod liver oil were unpredictable at best. Then there
was the oil's nasty source: pharmacy historian Aaron Ihde noted
that science-oriented doctors and pharmacists on the cusp of the
age of modern biomedicine were increasingly interested in new
chemicals and microbiology and came to disparage the oil as
"*Dreckapotheke*"—animal secretions or excrement used as medicine.[36] Finally, cod liver oil's physical action in the body was not
understood, and would not be until the 1920s, and the cures it was
credited with seemed inconsistent and unlikely to resolve in the

kind of clear and specific cause-and-effect relationship that was modern biomedicine's quarry.

If so, why then did doctors continue to prescribe it? The answer is framed in two statements about cod liver oil quoted above: Robert Darbey's declaration in 1782, "Except bark, opium, and mercury, I believe no one medicine in the material medica is likely to be of greater service"; and Robert Hunter's conclusion in 1857, "It is simply good nourishment and nothing more."[37] The seventy-five years these quotes span saw monumental changes in "scientific medicine," and cod liver oil's role shifted with the changing times. The now-infamous bleedings, pukings, and purgings of Darbey's age—all the depletive therapies of "heroic medicine"—were an aggressive but rationalist approach to what were essentially humoral theories. The therapies were grounded in arcane but erudite medical theories and had predictable effects, casting the physician as harsh puppet master to his patient's symptoms. To Darbey and those who had adopted it for rickets, cod liver oil fit perfectly the scientific doctor's arsenal, perhaps next to arsenicals (though, truth be told, the only "heroic" aspect of its administration was enduring its taste).

The trouble was, as the nineteenth century wore on, it became increasingly obvious that for all their dramatic effects, heroic therapies were largely impotent against the major causes of illness and death—undercutting popular acceptance of very unpopular treatments. As confidence in the old system of heroic depletive therapies crumbled, uncertainty about what should replace them cast a pall of doubt over medical science. But the rules were changing. The idea of self-limiting diseases took hold—why put patients through depletive therapies for those diseases that would run their course regardless? Instead, work to enhance the body's ability to ride out the illness. Prescribe rest, nutrition, hydration, fresh air. Build the body up. And the humble cod liver oil was a splendid tonic for enhancing the patient's constitution. Of course, as we have seen, the science of rickets in the twentieth century would once more transform cod liver oil into a miraculous specific against rickets. But as a historical survey from that later era concluded, doctors continued to prescribe it "because, and only because, it is an easily digested and easily assimilated fat."[38] And prescribe it they did. By the end of the century Manchester Children's Hospital was dispensing almost 2,500 gallons each year for rickets and tuberculosis.[39]

:   :

One notable consequence of the mid-nineteenth century's therapeutic revolution and the turn away from physician-driven heroic medicines was a frenzied market for home remedies, alternative health systems, and flat-out quackery, a cataract of drug makers with their hyperbolic claims offering the antidote to therapeutic nihilism. Cod liver merchants fell right in: a pharmacy trade journal from the early twentieth century listed twenty-one unique companies—from Baker's to Wilbor's—marketing cod liver oil.[40] Their ads took full advantage of the oil's full history, from its origins as a time-honored folk remedy to its place in the modern physician's tool kit. Scott's Emulsion was one of the most recognizable purveyors.[41] A typical Scott's ad from 1889 focused on the product's potency "as a flesh producer," promising to cure "consumption, scrofula, bronchitis, coughs and colds, and all forms of wasting diseases." Scott distinguished itself from the competition with its flavor: "As Palatable as Milk," thanks to its mix of "hypophosphites of lime and soda."[42] Scott's iconic trademark, a fisherman carrying a cod as long as he is tall, appeared on Scott's products and advertisements from the 1880s to the twenty-first century. Ludovicus Josephus De Jongh took a slightly different tack. In 1849, when the scientific inquiry into cod liver oil was still going strong, the Dutch physician published an influential analysis of *The Three Kinds of Cod Liver Oil*.[43] He clearly preferred one kind and leaned on his scientific credibility to promote it. "Dr. De Jongh's Light-Brown Cod Liver Oil" quickly became a leading product in Europe and the United States, its popularity buoyed, no doubt, by what can safely be judged a 210-page infomercial, his 1854 *Cod Liver Oil: Causes of its Frequent Inefficacy and Means of*

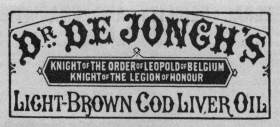

Figure 5.2 Dr. De Jongh's Light-Brown Cod Liver Oil was one of dozens of brands, but one of the most famous, mostly because of the tireless efforts of the Dutch medical man Dr. Ludovicus Josephus De Jongh. Leaflet from author's collection.

Removing the Same; with Remarks upon the Superiority of the Light Brown over the Pale Oil, Directions for its Use and Cases in which the Oil has been used with the Greatest Effect.[44] Those uses and cases were many, and the claims audacious.

While rickets appeared in many ads, conditions other than rickets got greater attention—none more than "consumption," which typically headed the long list of conditions any given cod liver oil claimed to treat. Wilbor's Compound of Cod Liver Oil and Lime pitched directly "To the Consumptive," offering to mail curious readers testimonials to its effect. The word, CONSUMPTION, set in large, all-caps type, formed the headline for an 1880 ad for Osmun's Cod Liver Oil and Lacto-Phosphate of Lime. Some of Osmun's ads mentioned rickets, though typically at the end of the usual catalog of complaints.[45] In most advertisements, the target conditions were less important than the oil's power to build the customer's constitution in order to fend off disease. Scott's promised their Emulsion "creates appetite for food, strengthens the nervous system, and builds up the body."[46]

With the endless hoopla of its marketeers, its status as a traditional home remedy, and the ups and downs of scientific interest, cod liver oil built a sizable footprint in popular culture, appearing in prescriptive literature of all sorts, as well as memoirs, poetry, and song, none more famous than the adaptation of a British broadside ballad by Newfoundland folk singer Johnny Burke (1851–1930). "Cod Liver Oil" tells the story of a despondent man who is ready to give up on life because his wife wants nothing but to drink cod liver oil. His house, he cries, "resembles a big doctor's shop, With bottles and bottles from bottom to top." The chorus shows how familiar some brands of cod liver oil had become, even as it gently ribs their ubiquitous advertisements:

> Oh doctor, dear doctor, oh doctor De Jongh
> Your cod liver oil is so pure and so strong
> I'm afraid of me life, I'll go down in the soil
> If me wife don't stop drinkin' your cod liver oil[47]

And then there are the jokes. Bad jokes: "Whether you believe the story about a bluebird that caught a cold and had to take cod liver oil, at all events it is a good story for the chilled wren." Political jokes, like this from the late antebellum years: "The Democracy of the country is in the last stages of galloping consumption, even Cod Liver Oil won't save it!" Obscure jokes: "A Delaware man has been taking cod liver oil for four years to cure the consumption, and has just found out that he never had any consumption. He is the maddest man in America, and his children haven't said 'boo' for a week."[48] Jokes about cod liver oil's taste: "Dr. Antonin Martin says that the flavor of cod liver oil may be changed to the delightful one of fresh oysters, if the patient will drink a large glass of water poured from a vessel in which nails have been allowed to rust"; or "the pleasantest way to take cod-liver oil is to fatten pigeons with it, and then eat the pigeons."[49]

Cod liver oil's well-deserved reputation for fishy, and some-times foul, taste was nothing funny to those doctors who wanted to heal their patients with it or to those who wanted to make their fortunes selling it, a problem succinctly stated by the purveyor of a cod liver candy: "Every one knows the virtues of Cod Liver Oil, but it is like a 'sealed book' to thousands of suffering invalids, on account of its repulsive taste and smell."[50] At the Manchester Infirmary in 1782, Robert Darbey admitted that "its nauseous smell and taste" were so bad that "many delicate stomachs can-not take it." Percival recommended rendering the oil "much less offensive" by a mixture of mint and various compounds; or, since "the oil is probably more efficacious in its original form," just downing the dose straight, with a chaser of "some acidulous li-quor" to "cleanse the mouth and gullet."[51] De Jongh also offered suggestions for masking the flavor, but he argued that it was "cer-tainly injudicious to aim too much at pleasing the palates of our patients." He preferred the more pungent "light brown" grade over the light, milder "pale oil," because, while a "well-tasted cod liver oil may be prepared," he worried that it might "be found wanting in" the most active ingredients.[52]

Many merchants ignored De Jongh's advice to accept cod liver oil's fishy taste as a feature instead of a bug. Some promoted mild-flavored oils.[53] Others followed the lead of Scott's Emulsion, enhancing the natural oil with other flavors and medicinal ingre-dients. Still others did their best to hide the flavor completely, as in the case of Bliss's Compound Cod Liver Oil Candy, manufac-tured in antebellum Massachusetts. Bliss's marketed in both the North and the South, apparently with great success. In 1851, the candy's inventor received a letter from a druggist "down South." The candy was selling "first rate, and was doing a great deal of good," and the apothecary wanted to replenish his supplies—but with one condition: "If the manufacturer was an abolitionist, the writer could receive no more of it." Bliss immediately shipped "a large consignment to his Southern customer." As for proving his political affiliations, Bliss referred the southerner to "the Town Clerk and Postmaster of Springfield."[54]

Cod liver oil companies continued to promote their product as a general "constitutional" into the early twentieth century, even as the pace of scientific inquiry into rickets jumped into high gear, and as evidence mounted in favor of cod liver oil's reputation as a specific against rickets. The previous chapter showed the critical importance of cod liver oil in the community studies, laboratory experiments, and clinical trials that established the modern science of vitamin D. The community studies often ended with a call for making cod liver oil freely available to the poor, while the clinical studies' findings would seem to encourage modern hospitals to embrace it as enthusiastically as the Manchester Children's Hospital with its annual consumption of 2,500 gallons. It seemed reasonable that through marketing and medicine, combining the power of the Advertising Age with the Golden Age of Medicine, America might realize the goal of eradicating rickets with cod liver oil and sunshine. But anything resembling universal supplementation—distributing the oil, getting parents to use it properly, and, always, overcoming the taste problem—was going to be a heavy lift. With the arrival of the Steenbock process for turning everything from dead skin to sheep sweat into a source of vitamin D, lowly cod liver oil's days would seem to be numbered.

: :

In the spring of 1929, the *New York Times* ran the headline "New Drug as Rickets Cure" over a story about the Steenbock process. In a speech at Cooper Union the night before, Alfred Hess had described vitamin D–enriched foods as "equivalent to taking ultraviolet rays internally." It had taken some time for the full potential of irradiating foods to be recognized. Even Hess, who had independently discovered what came to be known as the "Steenbock process," had initially been underwhelmed, predicting that its highest value might be in producing a cod liver oil substitute in the event of shortages.[55] In 1927, the *New York Times* had given page-one coverage to Steenbock's promise to turn his patent over to the University of Wisconsin but downplayed the "new body building process" at the heart of that patent. "Patents have been applied for," Steenbock told the press, but the process was slow, and he expected "there will probably be considerable delay in their being granted, so that any action looking toward production under my invention is yet distant."[56] On the contrary, pharmaceutical companies had been courting Steenbock since he first published about his process, and within a year of securing his patent, he and the Wisconsin Alumni Research Fund (WARF) had made advance licensing arrangements with five major pharmaceutical companies to manufacture Viosterol.[57]

Other parties hoped to jump on the manufactured-sun bandwagon, but at first Steenbock limited licenses to products with clear dietary or medicinal value—yes to yeast, but no to chewing gum.[58] Controlling licensing was at least in part motivated by Steenbock's desire to protect the public from fraudsters. Not long after Steenbock announced the irradiation process, one such huckster set up a window-front mechanism that passed a flowing film of cotton-seed oil under a run-of-the-mill UV lamp before capturing the ostensibly vitaminized liquid into bottles. Another merchant offered a lamp with which customers could irradiate milk in their own homes.[59]

The Steenbock process promised a simple, flexible tool to provide a bottomless source of vitamin D. But how should this resource be used: as a stand-alone medication or as a means of fortifying foods? Which foods? The energetic support from the evaporated milk manufacturers featured in the previous chapter suggests the answer. They saw vitamin D fortification as a means to cast their product as a "natural" foundation of infant nutrition and paid for the science that would make that case. In addition to the pharmaceutical companies, WARF licensed grain companies Quaker Oats and General Mills; Standard Brands received a license to irradiate Fleishmann's yeast. Licenses to the dairy interests were plentiful, doled out with an eye toward maximizing geographic distribution of vitamin D milk.[60]

Over time Steenbock dropped what he came to call "an ultra-idealistic attitude" and eased off strict limits on licensees. Among the licensing requests rejected in 1929 was Anheuser Busch, who wanted to fortify its beer. WARF and Steenbock held fast to restricting licenses to "the most important food products."[61] Seven years later things had changed: the Joseph Schlitz Brewing Company began marketing Sunshine Vitamin D Beer in 1936. "Beer is good for you," one ad gushed, "but Schlitz, with Sunshine Vitamin D, is extra good for you."[62] The beer that made Milwaukee famous did not have a license for the University of Wisconsin's irradiation process, but their brewer's yeast, which was fortified "directly by the ultra-violet rays of the sun" did.[63] When the trade journal *Modern Brewer* published an article on the benefits of fortified beer, the WARF's board considered taking action. Steenbock counseled restraint—near the end of their patent's lifetime, the foundation's status as "protecting agent" had been undermined by competing methods for providing vitamin D. The ground had shifted. Steenbock concluded that "if the public should demand vitamin D in its beers, there is no reason why the foundation should not provide it—because it may do some good and it most certainly will not do any harm."[64]

The example of margarine provides a notable exception to WARF easing "ultra-idealistic" licensing policies. Margarine would seem a prime candidate for vitamin D fortification, since butter was a natural source of vitamin A, but WARF and Steenbock held the line against oleo for the life of their patents. Their hard line is consistent with the university's long-running hostility to oleomargarine, long perceived as a threat to Wisconsin dairy industries, labeled a "greasy counterfeit" since the 1880s, subjected to punitive federal taxes, and held up as an infamous example in the pure foods campaigns burning through the dairy states in the Gilded Age.[65] Protecting Wisconsin's dairy farmers was critical for Steenbock; in fact, according to one of his colleagues, the "primary reason for securing the patent was so that license might be withheld from the oleo interests, to protect Wisconsin's dairy interests."[66] Although butter still dominated the market for spreads in the 1920s, butter shortages in the Great War had temporarily boosted oleo's market share, and the cheaper spread was shaking

Figure 5.3 As Americans learned about vitamin D and its value in maintaining health in both adults and children, pharmaceutical and food companies pitched all manner of vitamin D medicines and vitamin-enriched foods. Joseph Schlitz Brewing got in on the act with "Schlitz with Sunshine Vitamin D." *Esquire*, November 1936, 175.

off its reputation as "poor man's butter," with well-established brands, market organization, and coordinated advertising promoting margarine as the economical, modern spread.[67]

: :

Americans were not going to get their vitamin D from margarine or from beer.[68] But without a doubt, the biggest loser in the race to provide "the sunshine vitamin" was sunshine itself, whether in the form of natural sunlight or the artificial sun of UV lamps. Both had their committed champions, but neither could overcome the realities of how American cities and homes were being built, and of the powerful appeal of fortifying foods as a shortcut to the sun's goodness.

Sunshine's chances seemed a lot better thirty years earlier. Medical science at the start of the twentieth century vigorously pursued the power of the sun to heal; so too did American homeowners and city planners look to the sun to promote and preserve health. Fresh-air sanitaria and hospital rooftop wards were prominent sites of cure, and their preventive counterparts could be found in front porches, sunrooms, and sleeping porches in private homes; in rooftop classrooms; and in the general urban environment where city ordinances mandated minimum standards for light and air in tenements, wider streets, and building setbacks to bring light into city streetscapes. Faith in the healing power of sunlight played a large part in Progressive Era campaigns for parks and safe outdoor recreation for city children. Advocates for sunnier cities worried about city children of all classes, but often focused on the poor, whose "only playground is the gutter," and whose "only view of God's sunlight is through its reeking vapors."[69]

But these measures were not up to the challenge of the modern city, which remained dark, or grew darker, as building density and height increased. Despite smoke abatement campaigns in many cities in the early twentieth century, urban air continued to grow thicker with coal smoke and other pollutants.[70] The new knowledge of UV light and the importance of vitamin D heightened parents' concerns that their children, "screened from the shorter rays of sunlight by clothes, window glass, fog, and smoke," were endangered by a lack of sunlight. Many turned to new technologies to enhance exposure to the sun or provide a replacement for it.[71] The late 1920s to the 1940s saw energetic marketing of sunlamps for the home. Advertisements targeting middle-class consumers often emphasized the difficulty average families had providing their children with adequate sunshine. "How fortunate those children whose parents have leisure to follow the sun," declared a 1929 ad for the Hanovia Alpine Sun Lamp. Another ad piled on the class comparison: "Robust, sun-tanned, 'light-blessed' children . . . are found rarely these days, except in homes unusually well situated climatically." The Alpine Sun Lamp could, at the turn of a switch, provide "radiant health" to "children that stay at home," at a cost any parents would gladly pay "for such peace of mind."[72]

A less successful product for bringing UV light into the home was window glass engineered to pass the full spectrum of sunlight, marketed by Vita Glass and several other manufacturers. Previously, filtering UV light had been seen as a positive good, in order to protect drapery and furnishings from fading. But in the 1920s, the science of UV light and vitamin D changed the equation, pitting the colorfastness of a homeowner's furnishings against potential danger to their children. "Doctors know the value of bathing the body in the powerful health-laden light of day," one Vita Glass ad noted, "but it would shock most people to learn that real daylight seldom enters their homes." Yes, the special glass was a bit more costly, but, their ads promised, Vita Glass would save money otherwise spent on "expensive patent medicines and special foods."[73] Despite heavy advertising and early enthusiasm in the late 1920s, by the mid-1930s manufacturers had large stockpiles of unsold UV-friendly glass.[74] America had not soured on vitamin D, but as with solaria, sunrooms, rooftop classrooms, and UV lamps, the tide was turning in favor of taking the sunshine vitamin by the spoonful or the glass.

A quick survey of twenty-first-century consumer culture confirms that the appeal of bringing more of the sun's rays into the

Bathed in Living Light!

Doctors know the value of bathing the body in the powerful health-laden light of day, but it would shock most people to learn that real daylight seldom enters their homes.

Whole light which brings health and cures disease cannot pass through ordinary window glass.

Yet this is actually true, for real light includes the invisible health rays that bronze the skin, enrich the blood and increase the vitality. These vital *invisible* rays cannot pass through ordinary window glass, no matter how much light pours into your rooms.

But "Vita" Glass—the new inexpensive window glass—lets through the health-giving rays which ordinary window glass shuts out entirely. "Vita" Glass floods your rooms with living light and brings outdoor health to you and your children in your own home.

Take out the ordinary glass from your windows and put in "Vita" Glass. Get the tan of health in your cheeks and build up your vitality to resist colds, influenza and more serious ailments during the colder weather. Let "Vita" Glass save the money you now spend on expensive patent medicines and special foods.

"Vita" Glass is already used in many Hospitals, Schools, Offices, Factories and Private Houses. Supplies may be obtained from your local Glass Merchants, Plumbers, Glaziers or Builders. Send the coupon now for the wonderful story of "Vita" Glass.

"VITA" GLASS MARKETING BOARD, 24, ALDWYCH HOUSE, W.C.2
Please send me your booklet, "Health through Your Windows"

NAME _____

ADDRESS _____

V59 _____

"VITA" GLASS

"Vita" is the registered Trade Mark of Pilkington Brothers Limited, St. Helens

Before replying to advertisements refer to page 94.   97

Figure 5.4 Some glass manufacturers produced windows that did not filter UV light. British glass maker Pilkington promised their "Vita" Glass "floods your rooms with living light and brings outdoor health to you and your children in your own home." *Good Housekeeping,* January 1929, 97.

home or mimicking them with electricity was short lived. Premier home window suppliers today assure potential customers once more that their glass *filters out* "harmful" UV light; their promotional literature might mention health, but they emphasize protecting "home assets"—fabric protection is paramount once more.[75] UV lamps may be a crucial ally in fighting disease and disinfecting surfaces and spaces in hospitals and in the food industry, but families no longer buy sunlamps to protect their growing children from rickets. Instead, as even casual examination of the products on the dairy aisle in any American grocery store confirms, the United States took the "universal" route, fortifying most milk sold in the country with a recommended dose of the sunshine vitamin.

It might seem that with the rise of fortified milk, cod liver oil would find its way to the dustheap along with the discarded UV lamps and Vita Glass. But the stinky oil would still have a few good years—some of its best, in fact—even as America elevated vitamin D milk to its erroneous status as the "natural" source for the sunshine vitamin.

Improving upon Perfection

Modern Vitamin D Fortification

From the discovery of vitamin D, the notion of fortifying milk to prevent rickets had strong support from health professionals. "There is a plethora of remedies for rickets," researcher Albert Hess told the *New York Times* in 1933; "our difficulty is in selecting the best one." If we were going to undertake a program of rickets prevention through universal food fortification, he argued, milk was "an automatic method, since every child is fed on milk," so the added vitamin D would be a fitting adjuvant to "the calcium and phosphorus elements in milk."[1] Some physicians agreed with Hess's logic but saw *universal* fortification as unnecessary overkill. Later, after the question had been settled in favor of universal fortification, Johns Hopkins pediatrician Edwards A. Park recalled that he had not considered rickets as "a sufficient menace to warrant all milk being reinforced with vitamin D." Equally effective he believed, would have been "to have a special milk fortified with vitamin D" for children only.[2]

The American dairy industry, not surprisingly, agreed with Hess, not Park. From the opening of the century dairies had struggled to rescue milk from its oft-deserved reputation as a suspect and often deadly product.[3] By 1930 their campaigns had largely succeeded in lifting milk to its perennial status as "Nature's Perfect Food."[4] But the dairy industry had to guard jealously that fragile reputation; associating itself with the new food science of vitamins and the New Public Health promised to improve upon perfection. Milk producers took an active role in making vitamin D milk an indispensable tool in the quest to prevent rickets. In addition to participating in clinical and community trials financed by WARF, several large dairies secured approval for fortification from the American Medical Association, through its Committee on Foods. And they made their voices heard in the professional literature as well. The research director for The Dry Milk Company penned an editorial in the *American Journal of Public Health.* "Since milk is the sole or major article of diet of every child during the age of greatest susceptibility to rickets," he argued, irradiated milk offered "a new type of prophylaxis, simple in application, economical, and entirely free from the inherent handicaps involved in the use of the better known specifics," referring of course, to cod liver oil in its various guises.[5]

And so, the question became how best to transform milk, which by itself contains little or no vitamin D, into a reliable delivery vehicle for the sunshine vitamin. Steenbock, Hess, and others had shown that UV light's flexibility offered many approaches to adding vitamin D to dairy, and eager researchers tried just about all of them. Today, milk is fortified at the dairy processing plant by adding a measured, potent dose of vitamin D_3 (cholecalciferol) suspended in an oil emulsion. This method was clearly feasible in the 1930s, but regulatory and cultural resistance to "adulterating" milk stood in the way.[6] WARF, which administered the patents for the Steenbock process, actively discouraged dosing milk with an additive, wary of "the public attitude upon the question of the addition of foreign substances to milks" and recognizing "many statutes and ordinances prohibiting such a practice." Several pharmaceutical companies held patents from WARF to produce vitamin D concentrates for medicinal applications, but the foundation discouraged them from pursuing, even experimentally, "any process . . . wherein a foreign substance is added" to milk.[7] One company in New Jersey avoided WARF's patent and flaunted the old rules by fortifying their milk with Vitex, a vitamin D concentrate derived from cod liver oil via a process patented by Columbia University.[8] Vitex-fortified milk would be one of three competitors in the early commercial trials of vitamin D milk.

One way to add vitamin D without "adulterating" milk was to fortify the cow instead. Dairy researchers took two approaches to this noble goal. The first was to irradiate the cow (fortification *in uddero?*). As early as 1925, before Steenbock received his patent for irradiating foods, a team of researchers at the University of Maine's Agricultural Experiment Station in Orono tested this premise, sequestering four Holstein cows in a barn, exposing two subjects to ultraviolet light from electric lamps suspended three feet above their backs. Milk from the test subjects and the controls was fed to chickens with experimentally induced rickets. The milk from the irradiated cows had some effect on the rickety chickens but less than cod liver oil or exposure to UV light.[9] A larger study undertaken at the University of Pennsylvania, by much of the same team that we saw in the last chapter inducing rickets in infants at the Children's Hospital of Philadelphia, enrolled fifty cows. For nine months, their flanks and udders were bathed for fifteen minutes in UV light from high-intensity carbon arc lamps. Samples of the milk from the irradiated cows contained over four times the vitamin D as in the milk from control cows.[10] One study explored taking Bossie out of the equation, at least for nursing human infants. Alfred Hess fed some rachitic lab rats a diet of human milk from a woman who was "irradiated every other day by means of a mercury-vapor lamp." Hess found that indeed the irradiated woman's milk was an active antirachitic.[11]

The second approach, far simpler than turning cow stables into tanning salons, and a method WARF could countenance and that had patent licensees ready to try it, involved fortifying cows by feeding them a diet enriched with irradiated yeast. Although early clinical studies showed this so-called metabolized milk contained enough vitamin D to prevent rickets, in practice the vitamin content in the resulting milk was highly unpredictable.[12] Nonetheless, metabolized milk was the first fortified milk widely test-marketed, starting in 1931. Dairies in several states continued marketing

Figure 6.1 One of the ways dairies tried to fortify cow's milk with vitamin D was by exposing cows' flanks to ultraviolet light from powerful carbon arc lamps. John McK. Mitchell, John Eiman, Dorothy V. Whipple, and Joseph Stokes, "Protective Value for Infants of Various Types of Vitamin D Fortified Milk: A Preliminary Report," *American Journal of Public Health* 22, no. 12 (December 1932): 1220–29.

metabolized milk through the 1930s, though by 1939 only four thousand cows in American dairies were being fed irradiated yeast for this purpose.[13]

A far more reliable means of boosting milk's vitamin D content was to irradiate the milk itself. WARF worked with dairy equipment manufacturers to develop devices that passed a continuous thin stream of milk under intense UV lamps, activating the fat in the milk (7-dehydrocholesterol) to produce vitamin D3.[14] To the twenty-first-century reader, "irradiated milk" evokes the controversial modern practice of sterilizing foodstuffs by exposing them to gamma radiation. But in the pre-Hiroshima 1930s, faith in science and the miracles it offered was soaring. Dairies could confidently sell irradiation as safe *because* of its modernity and association with cutting-edge science: irradiated milk offered "over 8 years of clinical tests behind it" one ad promised. Far from downplaying the artificiality of the process, they boasted irradiated milk was *better* than nature. "Yes folks," a Borden ad assured, "we found a way to create as much Vitamin D in a few seconds as Old Sol could hope to give you in many days of direct sunlight;" irradiating with ultraviolet rays was "a method similar to the action of the sun's rays on the human body."[15] The technology was exciting: one dairy extended "a standing invitation" to the public to see "the only Irradiating equipment in Westchester," and one Chicago irradiator manufacturer exhibited a cutaway milk irradiator at the 1933–34 Century of Progress World's Fair.[16]

Irradiating milk allowed far more control over the potency of the vitamin boost, but it was highly technical. The dairy operator needed to monitor the intensity and duration of exposure: too little UV light, and the milk would not be adequately fortified; too much

Figure 6.2 For much of the 1930s, the most popular way for dairies to fortify milk with vitamin D was irradiation—passing the milk in a thin stream past a powerful ultraviolet lamp. At Chicago's Century of Progress Exhibition (1933–1934), visitors could inspect this cutaway of an industry-grade milk irradiator and read about the virtues of vitamin-D fortified milk. COP_17_0001_00011_007, Century of Progress Digital Image Collection, Special Collections and University Archives, University of Illinois at Chicago.

UV spoiled the milk's flavor, imparting a burnt taste.[17] The equipment for irradiating milk was costly to buy but inexpensive to operate. For large dairies that could absorb the start-up costs, irradiation was cheaper than other methods, contributing to its widespread acceptance. Direct irradiation would be the most-employed method in the opening years of the age of vitamin D milk.[18]

: :

Contrary to the myth that "in early 1930s essentially all milk in the United States" was enriched with vitamin D, the process of fortifying America's milk supply was a slow one.[19] Groceries in Boston, New York, and Philadelphia took the first small step in 1931, test marketing milk from cows fed irradiated yeast. The next year dairies in Detroit began marketing both irradiated milk and milk fortified with Vitex, the cod liver extract patented by Columbia University.[20] The pace of distribution picked up in cities in the Northeast as well as a few in the South and the West. But even after ten years, vitamin D milk was largely available only in urban areas, and less than 10 percent of the nation's milk was fortified. A 1947 dairy journal, praising fortified milk's "major significance" for milk distributors, boasted that "in many major markets vitamin D fortified milk constitutes from 30 to 60 per cent of the total fluid milk sales." Fortifying about half of the milk sold in "major markets" can hardly be considered universal penetration.[21]

The slow pace of adoption does not appear to have been due to popular or scientific resistance. A few doctors did raise concerns about possible overdosing of the vitamin or questioned whether the real motive was to benefit manufacturers exploiting the worries of "overanxious mothers." A California pharmacology professor dismissed fortifying milk as a "commercial enterprise . . . endeavoring to capitalize to the fullest the current popular vogue for vitamin D."[22] Some city boards of health needed to be convinced to allow "adulteration," but that rarely posed much of a hurdle, and only a few major cities demurred.[23] Chicago's Board of Health approved of vitaminizing milk in 1934. The New York

City Board of Health amended its sanitary code in 1935 to permit sale of milk vitaminized by any of three ways—by feeding cows irradiated yeast or by two other means that needed board approval: direct irradiation (by the Steenbock process), or the addition of a vitamin concentrate (initially by Vitex, a cod liver extract).[24]

From the mid-1930s, the dairy industry began promoting vitamin D milk in earnest, with print and radio advertisements.[25] In late May of 1935, just days short of the starting date set by the Board of Health, New York's Sheffield Farms announced: "Starting June first—you can give your children a better chance than you had." The copy noted but did not highlight that Sheffield milk was fortified with a flavorless extract from cod liver oil. A Borden's ad from the same day emphasized their choice of fortification methods, which they proudly included in their product's very name. "Be among the first to get the benefit of Borden's Irradiated Vitamin D milk." The first ads sought to introduce the new products and reassure the public of their safety and quality. "It is not an experiment" Borden's ad promised, "but a milk that has proved its worth in seven years of clinical tests with large groups of babies" and in test markets. Many early ads leaned on medical experts to win customers' trust: assuring that fortification was accepted by the American Medical Association's Committee on Foods or urging customers, "Ask your doctor about this special milk." One ad bragged that hundreds of their customers "are now taking Willow Brook Irradiated Vitamin D Milk . . . on the advice of their doctors."[26]

The early ads also tamped down concerns over cost and flavor. Dairies promised their fortified products would remain "within the reach of every community and every income bracket." Borden argued that the benefits of its Irradiated Vitamin D milk "are priceless—but the cost is only one cent more than a quart of Borden's Golden Crest Milk." As for assuring customers on the taste of the new milks, dairies that fortified with cod liver extract probably had a tougher sell. In 1935 customers were sure to associate cod liver oil with rickets prevention, vitamin D, *and* its famously bad flavor, so Sheffield emphasized that their method conveyed all the medicinal power of cod liver oil but "does not change the flavor of the milk in any way." Irradiated milk was more of an unknown, but Borden told customers that (implying but not directly criticizing the competing methods) "no foreign substance is added to the milk. Its flavor is not affected."[27]

As autumn approached, the ads began hyping the "sunshine vitamin" angle of fortified milk. Over the figure of a man executing a swan dive, one ad exhorted: "To keep summertime health, drink Borden's Irradiated Vitamin D milk." An ad from January took a harsher tone: "YOU CANNOT DEPEND ON THE WINTER SUN for all the Vitamin D your body needs." Promising "Summer in November," Sheffield assured that every bottle contained enough "sunshine vitamin" to "help in the formation and protection of sturdy bones . . . and help form and maintain strong teeth."[28]

Although the health benefits dairies touted for enriched milk went way beyond rickets prevention, rickets featured prominently in many ads, none more directly than a 1939 Sheffield ad featuring a girl wagging her finger at her doll: "You don't want to be bow-legged, do you?" or a Borden ad with the headline "LEGS tell the story!" accompanying a photo of a markedly bowlegged little girl playing a game with two straight-legged boys. This ad portrayed rickets as a common malady of a dark past: "A generation ago straight, sturdy legs were largely a matter of luck," but with irradiated milk, the whole family received "the precious 'sunshine' Vitamin D they need—and cannot get from their regular food, or the winter sun."[29]

Other ads were more playful. In early 1939, a Borden ad featured a cartoon cow merrily shaking hands with a walking letter D, the caption: "'D' Lighted to Meet You." The walking vitamin provides the ad's copy: "Maybe you didn't expect to see me popping up in these dark, wintery days."[30] In most of these ads the cow went nameless, though the copy might refer to Bossies or Flossies. But an ad in November used the name by which the Borden cow would forever be remembered. Under the image of an ardent, fan-fluttering cow leaning across the couch she shares with a cowering sun, the caption read: "Elsie's in the Moo-od for Love!" (Borden's

"YOU DON'T WANT TO BE BOW-LEGGED, DO YOU?"

Winter days need the help of extra vitamin D

RIGHT NOW, during these cold, wintry months, young bodies are demanding extra vitamin D. Strong, straight bones and well-shaped bodies depend on vitamin D. Milk is the finest natural source of calcium and phosphorus — minerals to help good growth. *Sheffield Vitamin D Milk* provides these minerals — *plus* enough vitamin D to put them to work in the body.

Don't wait another day. Order this extra benefit for your family. There are 400 U.S.P. units of vitamin D in every quart of Sheffield Vitamin D Milk. Enough to fill daily needs of a normal body. Phone for delivery.

SHEFFIELD *Sealect* VITAMIN D MILK

LEADERS IN QUALITY FOR 98 YEARS

¥0 Westmoreland Avenue
White Plains 9630

PROTECTED MILK FROM SELECTED FARMS

Figure 6.3 Sheffield Vitamin D milk got its "sunshine vitamin" from a cod liver oil extract. *Scarsdale Inquirer*, March 24, 1939.

"special irradiation process" was "wooing Old Sol's elusive Sunshine Vitamin D for the creamy richness of Borden's Milk"). From then on Elsie it was, and the increasingly anthropomorphic bovine became the face of modern pasteurized, homogenized, and vitamin-enriched milk.[31]

: :

Where did all this leave the much-maligned cod liver oil? Not in the dust, as one might assume. In fact, the early years of vitamin-enriched dairy products saw historic levels of cod liver oil consumption in the United States. Imports rose steadily through the 1920s, nearly doubled during the Great Depression, and reached a peak of 66 million pounds in 1939. Germany's invasion of Norway that year closed off the United States' main source; total imports in the war years never exceeded 22 million pounds. After the war, America lost its thirst for the oil: although imports ticked up from their wartime slump, they never exceeded even the modest numbers of the mid-1920s. This humble rally in turn faded by the late 1950s.[32] Cod liver oil's brief "golden age," coinciding as it did with the early years of heavy marketing of vitamin D milk, is worth exploring, not least because of the surprising role it would play in fighting rickets in the early years of "universal" milk fortification.

The first explanation for the cod liver boom in the interwar years is marketing. Cod liver distributors did not let their product take a back seat to synthetic vitamin D. Instead of the tiny text-driven advertisements buried in the back pages typical of earlier years, from the late 1920s big players such as Squibb and Scott's Emulsion placed quarter-page ads in *Good Housekeeping* and other major periodicals and newspapers, and ran full-page ads in more health-focused general periodicals like *Hygeia*.[33] These ads typically featured slick illustrations, often photographs of happy children sledding or playing in winter settings (wisely, they spent most of their ad budgets in the winter months when interest was greatest). "Little feet go marching bravely over the grim winter trail," an ad for McKesson's High Potency Cod Liver Oil began.

"D"LIGHTED TO MEET YOU

PERHAPS you don't recognize me without my usual coat of tan. And maybe you didn't expect to see me popping up these dark, wintery days. My monicker, too, is a bit misleading — "Sunshine Vitamin D" folks call me. Maybe you think I'm dependent on the sun for a living. Not any more! I've got a regular job now with the Borden people. Every day they put me into thousands of quarts of Borden's Irradiated Vitamin D Milk. This irradiation process of theirs is a marvel. In a couple of seconds it puts the same vitamins in the milk that old Sol gives when he shines all day. And all you do is drink Borden's Irradiated daily to get all the Sunshine Vitamin D you need. Why not start tomorrow? Just tell the Borden Man. Borden's Farm Products.

Figure 6.4 Borden began selling Borden's Irradiated Vitamin D Milk in 1939, promoting it with a series of newspaper ads featuring an anthropomorphized cartoon cow. Eventually the ads gave the cow a name, Elsie.

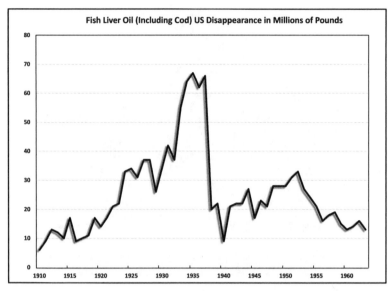

Figure 6.5 US Cod Liver Oil Consumption, 1912–1965. Economic Research Service, US Department of Agriculture, *Statistical Bulletin* No. 376: US Fats and Oils Statistics, 1909–65, Oilseeds, Oils and Meals, Animal Fats and Oils, Food Fats, Nonfood Fats.

Confident that readers knew the value of cod liver oil against winter's dangers, they bragged that their High Potency formula supplied "two and a half times the potency of vitamins A and D found in ordinary cod liver oil."[34]

Cod liver oil distributors continued their perennial quest to make tastier versions of their product. Scott's Emulsion had always pitched their modified elixir as more palatable than any plain cod liver oil. Now they had company: in 1929 Squibb started marketing "A New Pleasant Mint Flavor," Nyal sold a Wild Cherry and Cod Liver Compound, and another company marketed "Coco-Cod": "Cod Liver Oil Scientifically Combined With Egg Yolk, Rice, Yeast and Vegetables to which is added Pure Cocoa, imparting a delectable Chocolate Flavor."[35] Another approach was to pack the power of cod liver into pills. White's Cod Liver Oil Tablets claimed to have ended the "Cod Liver Oil Nightmare"—chillingly depicted in an ad featuring a terrified girl being attacked from above her bed by a sharp-toothed codfish.[36] Scott's marketed pills alongside their famous emulsion. Creative ways to mask or eliminate cod liver oil's taste persisted into the twenty-first century; if anything, the purveyors of cod liver oil have only gotten more creative in recent years, with gummies and a wide array of flavored oils.

And as in the previtamin era, cod liver oil ads emphasized the elixir's benefits beyond rickets, such as preventing colds, fortifying the constitution, and adding weight. "A Picture of Health," a 1934 ad for Scott's Emulsion, followed the popular winter-fun theme with two happy children sledding in the snow and testimonials on the subject of flesh-building and cold prevention. "My children used to be so thin and sickly," says one mother, until her doctor recommended Scott's.[37] A 1944 ad in the *New York Amsterdam* promised to "build stamina, energy and resistance."[38]

Of course, the best promotion for cod liver oil was its new standing in scientific medicine. Studies from the late 1910s proved the oil's power against rickets, but Elmer McCollum's 1922 identification of its active ingredient transformed the traditional elixir into a modern pharmaceutical. Oil merchants were quick to make the connection. Scott's Emulsion made a deep historical dive: "Since Viking Days," a December ad declared, "cod liver oil, now known to be exceptionally rich in the vitamines, has been a means of health and strength to tens of thousands." Squibb soon bragged that their cod liver oil was "richer in health-building vitamins than any other product known to science today!" These ads did not name Vitamin A and D but acknowledged the two "fat-soluble factors" McCollum had identified in cod liver oil and distinguished the body-building vitamin A from "the antirachitic vitamin"—"an infallible and complete protector against rickets."[39]

As soon as the labels "A" and "D" had been duly applied to the two "factors," advertisers adopted the nomenclature. A Squibb ad in late 1927 introduced readers to the new name: "A certain element—Vitamin D—is essential for the building of strong bones and sound teeth. And it is to be found in only two readily available sources. Direct sunshine and good cod-liver oil!" This ad, pitching Squibb's cod liver oil as "Bottled Sunshine," neatly tied most of the

COD LIVER OIL
NIGHTMARE...

NOW...fishy taste is unnecessary
—The good of Cod Liver Oil
is sealed in these tablets

FOR generations science has recognized the advantages of cod liver oil to the well-being of the growing child. Its one drawback was its repulsive fishy-fat flavor. Children hated it. Today's child can have her cod liver oil in the modernized pleasant way. Give *your* child White's Cod Liver Oil Concentrate Tablets.

These pleasant, candy-like tablets contain the health-promoting, resistance-building benefits of cod liver oil. Nauseating fishy fats have been removed. White's Cod Liver Oil Concentrate Tablets are a *natural* concentrate of vitamins A and D found in cod liver oil. Each tablet is always constant in its vitamin content—accurate in dosage because the vitamins are *sealed* against deterioration.

Accepted by the American Medical Association, Good Housekeeping Bureau and the American Dental Association.

FOR INFANTS, these tablets are easily crushed and mixed in milk, orange juice or tomato juice. No messy feedings.

FOR ADULTS, these tablets are the perfect, convenient way to build resistance. So easy to carry.

 COD LIVER OIL CONCENTRATE TABLETS

Figure 6.6 The arrival of manufactured vitamin D supplements did not immediately lead Americans to throw out their cod liver oil. On the contrary, cod liver oil sales skyrocketed in response to new scientific findings proving that the old elixir was a powerful rickets preventive. Drug companies sought to overcome the oil's "flavor problem" in creative ways, including White Laboratories' cod liver oil concentrate tablets, which promised to drive away the "nightmare" of cod liver oil. *Good Housekeeping*, December 1, 1933, p. 160.

new science of rickets into one salient quarter page of copy. First, it highlighted the power of x-rays to find invisible cases of rickets: "Your baby may *look* plump and rosy" and may be well-fed. "Yet the X-ray may show little bones and teeth developing soft and porous instead of solid and strong." Next it clarified the connection between vitamin D, UV light, and natural sunshine, warning that "weather, clothing and modern living conditions" make it "almost impossible" to get adequate vitamin D from the sun. Finally came the assurance of scientific consensus: "So authorities have come to depend for protection on what they call 'bottled sunshine'—cod liver oil!": prescribed by "leading baby specialists," and verified by laboratory tests of its potency.[40]

It made sense for the oil merchants to add new scientific rationales to their seasonal pitches for the traditional tonic. Popular magazines and newspapers had long reported on the major findings of the vitamin hunters. In addition, health-advice columnists regularly featured updates on the science, guidance on the benefits of cod liver oil, and, inevitably, advice on how to administer it—usually with some version of this physician-columnist's warning: "If you dislike the smell or taste of cod liver oil, never allow the child to know it." The writer celebrated the new flavored oils, reminiscing that "when I was a child, we were bribed to take the smelly tonic." A popular pamphlet from the US Children's Bureau echoed the advice to keep mother's dislike to herself, warning that "her facial expression" could "teach the baby not to like it. She must take it for granted that he will like it even if she does not."[41]

Government support for cod liver oil was not limited to Children's Bureau pamphlets. Historian Janet Golden found that many health agencies "preached the gospel of cod liver oil." At the federal level, child health advocates at the Children's Bureau and the Bureau of Indian Affairs made the oil a mainstay of their programs. Local hospitals and well-baby clinics continued the practice pioneered in studies by Hess and Eliot of handing out cod liver oil to the public, free or at cost. And the word got out: oral histories taken for the WPA showed that everyday citizens were well aware of the benefits of cod liver oil, knowledge they had gleaned from home economics classes and health clinics.[42]

Even as fortified milk started appearing in America's dairy aisles and milk trucks, and even though the AMA's Council on Foods had approved of milk as a vitamin delivery vehicle, many doctors preferred cod liver oil and other vitamin D supplements. American pediatricians surveyed in the late 1930s reported that even if vitamin D milk "were available, they would still prescribe cod liver oil, viosterol, etc." Some argued for reserving vitaminized foods "for infants who refuse to take cod-liver oil or with whom it disagrees," or worried that depending on vitamin D milk would "defeat the cod liver oil therapy that physicians had been encouraging for years."[43] Prescription vitamins kept physicians at the center of child health. Doctors wanted to monitor and adjust intake for their patients' needs and worried about children receiving unhealthy doses of vitamin D. One Mead's cod liver and Viosterol ad challenged the old advice to "tell your patients 'to get plenty of sunshine,'" since the modern environment and clothing styles "all militate against 'plenty of vitamin D,'" and "you cannot control the potency or measure the dosage" of sunlight. This emphasis on predictable dosage gradually steered doctors toward viosterol, or a fish liver oil fortified with viosterol to a reliable (and high) potency. Physicians in the mid-1930s were at sea when it came to the "correct" dosage for cod liver oil (a problem that still plagues them today). Some prescribed as little as one-tenth of a teaspoon per day, while others pushed for five full teaspoons per day. High-potency elixirs from reputable pharmaceutical companies took some of the guesswork out of the picture and reduced the problem, articulated by Hess, "that it is easy to prescribe cod liver oil but extremely difficult to be certain that it is taken regularly."[44]

Doctors got plenty of encouragement from pharmaceutical companies advertising cod liver oil and other supplements. In full-page ads in national and local medical journals, Mead Johnson and the other major players mixed Madison Avenue techniques with the language of modern biomedicine. Parke Davis sold a "Standardized Cod-Liver Oil" whose potency, they claimed, was scientifically monitored by the "first commercial laboratory to assay Cod Liver Oil for both vitamins A and D." Even the slickest ads might feature quotes from peer-reviewed medical literature, complete with AMA-style citations.[45] These ads favored fortified cod liver oil or full replacement with Viosterol. Mead Johnson encouraged doctors to consider "what so many physicians have found a successful practice": transferring cod liver patients over to Viosterol, at least in the summer, "when fat tolerance is lowest," since "the most squeamish patient can 'stomach' it without protest."[46]

Although some companies advertised in both popular and medical journals, others made a point of pitching exclusively to medical professionals. Mead's boasted their Oleum Percomorphum was "Ethically Marketed—Not Advertised to the Public," while White's assured druggists that their cod liver oil products were "ethical" because they were marketed by constant "face to face broadcasting" to doctors in every town "by our coast-to-coast network of trained detail men." Such discipline was in keeping with Progressive Era policies of the American Medical Association to rein in patent medicines by establishing standards and controlling advertisements, but Mead Johnson provided a reason that doctors might find more important: such policies assisted "in keeping pediatric cases in medical hands." It is easy to imagine how frustrating it must have been for the creative team who came up with the powerful Mead's "Shadow of Rickets" ad that their work could not find an audience beyond readers of professional journals.[47]

: :

History credits fortified milk with curing rickets, but perhaps lowly cod liver oil deserves a greater share of the glory. Modern common wisdom—expressed here by the NIH—holds that "the fortification of milk with vitamin D beginning in the 1930s has made rickets a rare disease in the United States." Some go further, saying "fortification of milk with vitamin D led to *eradication* of rickets in the United States" [emphasis mine].[48] While never eradicated, scattered evidence from the interwar years suggests that

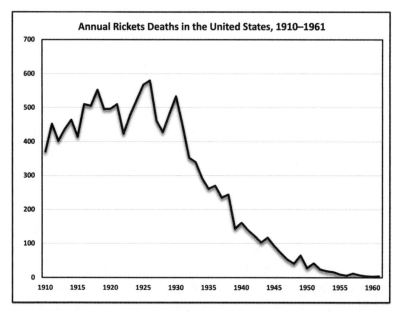

Figure 6.7 Deaths from rickets in the United States, 1910–1961. Weick, "A History of Rickets in the United States," *American Journal of Clinical Nutrition* 20, no. 11 (November 1967): 1234–41.

well before the age of vitaminized milk, old-time cures of sunshine and fish oil dramatically reduced the worst expressions of vitamin D deficiency.

For most of the twentieth century, the federal government collected annual records of deaths from rickets. In 1967, Sister Mary Theodora Weick, a nutritionist at Buffalo's Mercy Hospital, tabulated five decades of these death statistics for a brief historical survey of rickets in the *American Journal of Clinical Nutrition*.[49] And while they reveal a sobering total of almost fourteen thousand deaths, the data tell a convincing story of progress, in line with the optimism of the postwar years. Annual deaths attributed to rickets peaked at 580 in 1926, then plummeted to single digits by the late 1950s and remained low through the rest of the century—a very long tail to a vertiginous centuries-long graph.[50] But over half the decrease was achieved by 1934, before milk fortification was fully underway.

Data on nonfatal rickets are harder to come by, but they paint the same picture, if more impressionistically. For example, a 1936 review of thousands of examinations of preschool children in Chicago's infant welfare stations found "a decided decrease in rickets," beginning with a wobbling decline in the late 1920s, followed by a precipitous fall in the early thirties. Chicago Board of Health official Fred Tonney did not specifically credit cod liver oil for this progress, referring to all antirachitic agents "in sum total." But while he was optimistic about vitamin D milk's promise, only a quarter of the city's milk supply had been fortified, so "its potential usefulness toward solving the public health problem of rickets, cannot as yet be judged." It would seem that traditional antirachitics "in sum total" had brought about the change underway.[51]

Other experts were more explicit in their confidence in cod liver oil. Seeking to answer the question "Can Rickets Be Eliminated from the Large Cities?" a 1937 study in Newark examined one hundred infants who had been enrolled in the city's visiting

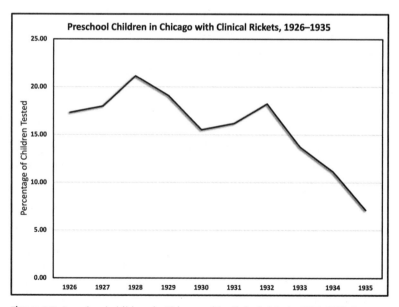

Figure 6.8 Preschool children in Chicago with clinical rickets, 1926–1935. From data in Tonney, "Vitamin D in Child Health," *American Journal of Public Health* 26, no. 7 (July 1936): 665–71.

nurse program since birth. The program involved monthly visits at "Baby-Keep-Well Stations" and periodic in-home nurse visits. Nurses provided cod liver oil free of charge, instructed mothers on administering it, and monitored their compliance. The one hundred selected infants, drawn from the poorest areas in the city, were given clinical and x-ray examinations for signs of rickets. About one-fourth showed some evidence of rickets, but only one was characterized as having "marked rickets." Cod liver oil had the "practical affect in reducing the incidence of rickets in the community to the point where it is not a clinical problem," answering the study's question and suggesting the answer to mine, about the relative contribution of cod liver oil. The authors concluded "that rickets can be eliminated from the infant population" with a consistent program relying on just cod liver oil.[52]

Such high hopes for cod liver oil had been common for decades. In 1923, before the Steenbock process made the magic bullet of universal vitamin D fortification possible, Edwards A. Park, the pediatrician who had trained Martha May Eliot, asserted that if pregnant women and infants were given decent diets, including cod liver oil, and if they spent "sufficient part of their time in the open air and sun . . . that rickets would be abolished from the earth."[53]

Twenty-five years after Park's observation, vanquishing rickets was still a distant goal, but cod liver oil's time in the sun was coming to an end. America had embraced vitamin D–enriched milk as the chief ally in that quest. Wartime shortages of cod liver oil put the traditional antirachitic at a disadvantage, especially in its competition with powerful pharmaceuticals like viosterol. More significant, though, was the dairy industry's growing commitment to universal vitamin D fortification, driven by consumer recognition of the need for vitamin D, the growing association of milk with vitamin D, and the sensibility of delivering in one popular product both calcium and the prohormone that helps the body process it— strong reasons vitamin D milk provided distributors a perennial marketing tool.[54]

If cod liver oil was sliding back into its traditional status as folk remedy and the butt of nostalgic jokes, most of the competing new-fangled methods for adding vitamin D to milk were becoming obsolete as well. "Metabolized" vitamin D milk, fortified by feeding cows irradiated yeast, never really caught on, and irradiated milk, despite its aura of ultramodern biotechnology and its dominant status, ultimately lost out to the approach that has dominated since the midcentury: the addition of a liquid concentrate of synthetic vitamin D3—by far the cheapest method, available to dairies large and small, and the method best suited for strict quality control.[55] The main reason liquid concentrates had not dominated from the start was skittishness about violating state and federal adulteration statutes. Any one of the pharmaceutical companies licensed to produce Viosterol could have provided dairies a reliable concentrate, had not WARF all but prohibited it. While it is true that the foundation developed a concentrate, UVO, to compete with the cod liver extract Vitex in markets where the adulteration ban had been set aside by local ordinances, WARF still focused on promoting irradiated milk.[56] By the late 1930s, the potential benefits of fortifying milk with a liquid concentrate swept aside Progressive Era objections to "adulterated" milk. Surely, the new logic asserted, modern dairies adding a concentrate of vitamin D was not to be compared with unscrupulous dairies of yesteryear polluting pure milk with chalk, sweeteners, or other adulterants. Vitaminization, one enthusiast insisted, "enhances the nutritional value of the milk, as opposed to the addition of an inferior substance which debases it." As early as 1934, the AMA's Committee on Foods had blessed this particular form of food adulteration.[57]

WARF's stance against "adulterating" milk with vitamin D concentrates was already softening by 1940, but in 1946 the foundation lost all ability to control the process, when the US District Court for the Northern District of Illinois voided all of the Steenbock vitamin D patents, ending a protracted antitrust suit. This was an ironic outcome to a process begun when WARF sued a vitamin manufacturer that was irradiating yeast without a license. The public and legal scrutiny from that case turned instead on the patent holder, casting WARF as a price-gouging monopoly. Eventually the government brought an antitrust suit against

WARF, charging that because of the foundation's monopolistic price policies, "persons who most needed Vitamin D to prevent and cure rickets were unable to get it."[58] From there, it was open season for any manufacturer to compete in the growing market for providing vitamin D concentrate to America's dairies.

: :

The vitamin D in today's fortified milk, D3, does not originate in sunlit cows or codfish. Nor is it synthesized by irradiating yeast as viosterol was—that makes vitamin D2 (ergocalciferol). Vitamin D3 is the form our bodies produce when we bask in the sun. It is the "D" in cod liver oil. D3 can be synthesized by exposing any animal tissue to ultraviolet light. As it turns out, the synthesized vitamin D3 added to almost all milk in America is the end product of a commodity flow extending over oceans and continents to the hides of millions of sheep—or more accurately, to their shorn wool. In the typical process, the vitamin manufacturer purifies lanolin extracted from sheep wool to produce crystalline cholesterol, which, when irradiated, produces vitamin D3. Lanolin from sheep is the preferred base, but the purification process is so thorough that in the end there is no way to determine whether the vitamin was derived from wool, cowhide, or pigskin.[59]

Viosterol, or synthetic vitamin D2, was WARF's flagship vitamin product. In the early years of synthesized vitamin D, researchers suspected (though they struggled to verify), that cod liver oil and irradiated milk were more effective against rickets than Viosterol and other plant-based vitamin D sources. It was only in 1937 that Vitamin D3 was identified as a unique substance, the same vitamin produced in the body. Verification of D3's greater efficacy followed soon thereafter, and, as a result, in the free-for-all years after the courts voided WARF's patents, the market quickly leaned toward vitamin D3 for human nutrition.[60] Today synthesizing vitamin D is a big business, dominated by multinational corporations. Some of the biggest names are familiar: BASF, Hoffmann-La Roche, DSM; others less familiar span the globe from Ireland to India, Germany to China.[61] Synthetic vitamin D is the engine of a multibillion-dollar global supplements market. With the renewed interest in vitamin D and rickets in recent decades, pediatricians, general practitioners, and geriatricians alike now monitor their patients' vitamin D levels as never before and prescribe supplements to young and old, pushing sales to over twenty times their levels at the turn of the century.[62] The bottles of vitamin D customers buy at the drugstore contain the same lanolin-derived oil infusion that goes into milk. And as American milk consumption continues its slow and steady decline, those golden capsules are an even more important source for vitamin D–hungry Americans.

In recent years, some health-conscious consumers chose to forgo the capsules in favor of getting their vitamin D from a more natural source, reinvigorating the market for cod liver oil, dormant since the 1940s. In the mid-1930s the United States imported over sixty million tons of cod liver oil each year. Toward the century's end, the elixir's popularity had plummeted over 90 percent.[63] Consumers seeking a natural alternative to capsules from big pharma quickly reversed this trend, and while cod liver oil never reclaimed its status as mainstay of infant and child vitamin supplementation (that honor goes to Flintstones chewables and the like), it remains a steady niche market, with dozens of brands competing in health food stores and online, selling bottled oil, whether "natural" unflavored or flavored with lemon, lemon-lime, orange, strawberry/banana, or mint. Shoppers can also opt for gelcaps, tablets, or chews in a rainbow of flavors.[64]

One inconvenient fact for purists turning to cod liver for a natural vitamin D boost is that the vitamin D in most commercial cod liver oils does not come from cod's livers but from the same source used in dairy—sheep's sweat. Unlike the oil produced in bygone days, from casks of putrefying entrails in fisheries, most oil on the market today is heavily processed to remove ocean pollutants that accumulate in *Gadus*'s busy livers. This purification destroys vitamin D, so manufacturers restore the vitamin content by adding the same synthesized cholecalciferol that dairies add to milk. This compromise has resulted in a new market for purists: cod liver oils processed without high temperatures or chemicals. The most popular of these upscale oils is known as "fermented cod liver oil"—a misnomer since the oil itself is not fermented but is released from cod livers that undergo fermentation, "such a gentle method," according to sales literature, "that it shields the natural antioxidants." A more recent competitor calling itself "Artisan Extra-Virgin Cod Liver Oil" extolls its chemical- and heat-free method of extraction. Competition between these new-age throwback supplements has at times been nasty, with extra virgin manufacturers warning of "rancid" fermented oils and accusations of adulteration or inadequate freedom from pollutants.

: :

The debates over the best cod liver oil in the third decade of the twenty-first century echo those of nearly two centuries earlier, when medical science sought to understand which of the many forms of this folk remedy offered the most promise to modern medicine. It seemed clear to a reviewer for the *Dublin Journal of Medical Science* in 1850 that too much debate over the best oil was counterproductive: "It is important that the cod liver oil should come into general use . . . and to this end it is well that the profession should be aware that there is no oil so bad as to have been proved to be without most valuable properties as a curative agent, nor so good as to have been demonstrated to be therapeutically more potent than the worst."[65]

Bad or good, natural or artificially fortified, modern cod liver oil represents the end point of a centuries-long attempt to control rickets, and despite its fall to the status of footnote (or worse—a joke) in that history, its historical reputation was burnished by its place in what came to be seen as the triumphal conquest of rickets, a conquest that was anything but complete.

A Long Shadow

By the middle of the twentieth century, a persistent myth had evolved about the end of rickets in America. "Has vitamin D milk accomplished what the doctors hoped it would?" asked *Today's Health* in 1952. "Yes indeed it has!" came the breathless reply.[1] Such confidence was not limited to the pages of popular magazines. In 1966 the *American Journal of Public Health* carried an editorial titled "The Disappearance of Rickets," penned by Harold E. Harrison, a pediatrician at Johns Hopkins University. Harrison was confident his peers would agree that rickets had disappeared and would not be "impressed too much" by that fact, having witnessed cures for polio and the advent of antibiotics "against most of the important bacterial infections." But Harrison, who with his wife, Helen Harrison, had studied rickets for over a generation, wanted to drive home the significance of rickets' disappearance. His readers, numbed by the miracles of vaccines and antibiotics, must understand that its elimination ranked "with the prevention of diarrheal diseases of infancy and of diphtheria as triumphs" of medicine and public health. Rickets, "once the most common affliction of childhood," had been successfully reduced to a status as "one of the rare diseases in this country and regarded as a medical curiosity in the major children's clinics."[2]

Some reports acknowledged that the cure was not total. "Rickets still is found with some frequency in some of our larger cities, *Today's Health* added to its celebration of vitamin D's victory, "especially among the dark skinned races"; but that couldn't detain the authors from concluding "it has almost disappeared in the country as a whole." With rickets among the "dark skinned races" thus removed from consideration, the professional health literature spoke in terms of "triumph," "conquest," and "cure." And the myth persisted, as demonstrated in a National Academy of Sciences web post from 2000 that proclaims: "By 1924, . . . children began consuming irradiated milk and bread and, seemingly overnight, the imminent threat of epidemic disease dwindled to a half-forgotten historical event."[3]

Half-forgotten, for certain. Dwindled, maybe. This chapter explores the gaps between the myth of eradication and the realities of a persistent if milder burden of vitamin D deficiency despite "universal" fortification of dairy and routine dosing with cod liver oil. The easing of concern in the last half of the twentieth century was a consequence of several factors, starting understandably with the truth at the heart of the myth: there had indeed been a clear and dramatic reduction in rickets, and that progress was but one of many steps in the general improvement in child health. But America's ejection of rickets from its anxiety closet was no doubt aided by the fact that the remaining burden of vitamin D deficiency fell heaviest on poor and minority populations, their risks, including obstetrical complications, deemed an inconvenient footnote in a triumphal history.

::

Rickets researchers in the early twentieth century had been driven by a common understanding that they were fighting a public health crisis of epidemic proportions. But with the discovery of vitamin D and the successes in deploying Steenbock's magic bullet of irradiated foodstuffs, priorities shifted. Hector DeLuca, the biochemist who succeeded Steenbock at the University of Wisconsin, recalled it this way: following the "monumental discoveries" of Steenbock's generation, "rickets disappeared as a major medical problem and the vitamin D research settled into a very quiescent state with only an occasional new discovery being made."[4]

Those researchers and clinicians decades earlier who enlisted cod liver oil and synthetic vitamin D to fight rickets needed only to know *that* it worked and find the best ways to use it. In the postwar years, researchers sought to understand *how* it worked and *what* it was. "The physiology of vitamin D," recalled Helen and Harold Harrison, remained a fascination for investigators, as "a biologic problem which continues to develop in complexity and interest even though the original mission of prevention and cure of rickets has been apparently accomplished."[5] Which is not to ignore earlier progress in theoretical research: in 1932 British chemist F. A. Askew identified D2 as a unique substance, followed in 1937 by the German chemist Adolf Windaus identifying D3 as the form of the vitamin produced naturally in the body. The biochemical complexities multiplied as a new generation of biochemists and physiologists untangled the chemical and metabolic processes at work in the body to transform vitamin D's building blocks into the powerhouse of calcium metabolism.[6]

On the clinical side, researchers, including Helen and Harold Harrison, turned to "interesting varieties of rickets" caused by conditions other than vitamin D deficiency. These comparatively rare forms included vitamin D–resistant and vitamin D–dependent rickets, caused by genetic mutations affecting phosphorus and calcium metabolism.[7] Another important project sought to prevent and cure rickets with a single massive dose of synthesized vitamin D, either orally or by injection. From one perspective this might seem like a way to move rickets management from the realm of public health firmly into the medical clinic. But a crucial application for this technology was situations where rickets posed a severe public health threat, as in much of Europe still recovering from the Second World War.[8] In professional journals of the 1950s and 1960s, studies of these less-common forms of rickets and new therapies vastly overshadowed reports on "traditional" vitamin

D–deficiency rickets. In the popular media, however, these subjects received scant attention.

As methods of adding vitamin D multiplied, concerns rose
that overlapping fortification schemes might supply too much of
a good thing. In wartime Britain, as in Germany, rickets prevention had been a national priority, driven by the experience of endemic rickets in the interwar years and, of course, national pride.
The Reich, determined that the "English Disease" never again be
a German disease, mounted an education program complete with
educational films and distributed potent vitamin D supplements,
including irradiated ergosterol manufactured by IG Farben under
a patent from WARF for the Steenbock process.[9] British supplementation efforts had been equally aggressive and continued after the war. In addition to fortifying all infant formula, cereals,
and margarine, the Ministry of Food further supplemented the
diets of poor Britons through the Welfare Foods Scheme, providing vitamin D–enriched evaporated milk and "National cod
liver oil compound" (cod liver oil mixed with acacia gum, syrup,
oil of bitter almond, and water). Without control over the means
of fortification, children might receive many times the dose required for rickets prevention, enough to cause hypercalcemia—
excess calcification in both bone and other body tissues. And in
fact, by the early 1950s, British doctors started reporting scattered cases of hypercalcemia—hardly an epidemic but concerning
enough to prompt the Ministry of Health to cut by half the vitamin D supplied through the Scheme and to recommend British
food manufacturers similarly cut back on vitamin D in foods sold
to the general public. Quickly the reported cases of hypercalcemia
dropped by half, without an increase in rickets diagnoses.[10]

This attention to rickets prevention in Europe before and after the Second World War helped reinforce the myth of America's
victory over rickets. Just as their grandparents had in the early
1920s, American parents of the late forties invoked the "starving
children of Europe" to tell their children to finish the food on their
plates. The generation between the wars had raised their children
in an America racked with home-grown poverty, crushing rates of

childhood hunger, and daily reminders of rickets and other deficiency diseases. But those hardship-tempered children had licked Old Man Depression as thoroughly as they whipped the Axis. For that generation, rickets in America must have seemed to have gone the way of the bread line and the fascist. America could now stand tall and assume the role of the world's policeman—and lifeguard.

"There is one grim and tragic battle of the Second World War which has never ended," a 1947 *New York Times* editorial proclaimed, "the battle of the children." Tens of millions of children in Europe and Asia faced starvation and disease, and the US, through the UN and other government agencies, needed to step in and protect children—the "Seeds of Destiny," as an American propaganda film of the time described them, "the human raw material of each shattered nation's tomorrow."[11] America's self-image was not one of a shattered nation, but one emerging from economic and military travail more powerful than ever, standing tall on strong limbs kept straight by vitamin D. Finally, in stark contrast to that image, the grim reality of polio took some of the focus from rickets, though the "conquest of rickets" through biomedical discovery and successful search for a cure may have encouraged the search for a similar miracle for polio.

::

Despite the progress revealed in the historical death statistics Sister Mary Theodora Weick highlighted in her brief 1967 survey of "The History of Rickets in the United States," she was not affirming America's triumph over rickets.[12] Weick acknowledged that "much has been accomplished," but for every rickets death, an unknown number of children had clinical or subclinical rickets. For a sense of that number, Weick presented statistics showing that from 1956 to 1960 American pediatric hospitals admitted 843 patients for rickets.[13] But hospitalization data, sparse as they were, still only captured the most severe cases. She urged that "rickets should be classified as a reportable disease and that all infants should undergo frequent examinations in order to detect it."

Weick complained that "as late as 1960, one State Health Officer said that rickets was not differentiated from other diseases . . . but the after effects were treated in Crippled Children's Clinics!"[14]

Those clinics had evolved over the twentieth century, growing from a patchwork of local institutions such as New York's Hospital for the Ruptured and Crippled, into a loose network of hospitals and local and state-run clinics providing short- and long-term care for disabled children. In a survey of the field conducted in 1912, the printer, typographer, and disabilities activist Douglas C. McMurtrie noted the ad hoc nature of the nation's response but assured his readers that the United States had "a great many more institutions for the care of cripples than is generally realized." He looked forward to some "national congress for cripple-care" that would exchange information and methods.[15] In effect, the US Children's Bureau's Crippled Children Program, authorized by the Social Security Act of 1935, established that "national congress." Through this program, all states received funding and guidance for programs to help children with disabilities. In turn, the Children's Bureau collected detailed statistics from each state's programs.[16] These data show that by the 1960s rickets and its aftereffects were still a persistent problem, even if other conditions, from polio to cerebral palsy, were far more numerous. Still, the numbers are sobering, with the Bureau-sponsored programs reporting from one to two thousand rickets cases each year, and upwards of ten thousand children treated for bowlegs or knock knees. Death statistics suggested rickets had been beaten. Falling admissions for treatment of rickets to regular pediatric hospitals, roughly a hundred or so each year by the 1960s, likewise signaled solid progress. But those thousands of boys and girls being fitted for corrective braces or undergoing corrective surgery to straighten legs curved by rickets tell a different story. Or, rather, they tell only part of the story: they were the ones whose conditions were severe enough, or whose families' financial resources were sufficient—to secure them a bed and a surgeon. For each of these "lucky" children, an unknown number went untreated.

A haunting report by a nurse-in-training in East Harlem gets at the odds against children with rickets receiving the care they needed—that is, of becoming a statistic, a patient whose condition and treatment are recorded, instead of remaining among the uncounted, sometimes dismissed as the "invisible" sufferers. In 1968, Sister Michael Joyce, a student at Boston College School of Nursing, worked through the summer before her senior year at the Little Sisters of the Assumption Home Health Agency.[17] She was assigned to care for "Mr. and Mrs. G." and their six children, aged from eleven months to five years. The family had recently moved from a farm in the South to a large apartment in a condemned tenement in East Harlem. After a succession of hospitalizations (for rat poisoning, head trauma, and dehydration) the family was on the radar of "a variety of health and social agencies," including the Society for the Prevention of Cruelty to Children, who referred the family to the Little Sisters. Joyce took the lead in devising and initiating a five-point intervention: providing child guidance, finding adequate and workable health care, improving nutritional status and "family relationships," and obtaining safe and stable housing. Their youngest, a severely malnourished ten-month-old, had the most pressing health needs, but his brother, Sammy Jr., ran a close second.[18] Sammy and his twin sister had both tested positive for tuberculosis, and Sammy was severely bowlegged. "His short brown legs formed an almost perfect 'O,'" Joyce reported, and his gait reminded her "of a monkey on the run." Sammy's mother assured Joyce that Sammy would "grow out of his crooked legs, like his grandfather," but Joyce added finding an orthopedist to the intervention program.[19]

The *American Journal of Nursing* published Sister Michael Joyce's story, "Assignment: East Harlem." It is a rescue story wrapped in a jeremiad. After two and a half months of daily home visits, each three to four hours, longer when necessary, and after ongoing efforts of a support team led by Joyce, the "G" family had met many of the initial goals. The two oldest girls were attending a nursery school, and the mother had "begun to assume responsibility for feeding the children regularly." The baby, Tiny, had gained eight pounds; all the children were enrolled at a family health care program in a nearby city hospital; and Sammy "was set to begin a course of treatment in an orthopedic hospital, with the hope that some correction of his deformed limbs may ensue." Joyce makes clear she is not the hero in the rescue drama, but a manager of a multiphased set of systemic and community-wide responses, the only approach that meets the family's multiphased and systemic needs, needs she introduces with an extended quote from Michael Harrington's *The Other America*: "Being poor is not one aspect of a person's life in this country, it is his life." It's a passage that includes Harrington's most illiberal notion, that "poverty is a culture." But Joyce quotes that passage to get to her thesis, which gently corrects Harrington's seeming determinism. She continues quoting Harrington. Poor families "are, in the language of sociology, 'multiple-problem families.' Each disability is the more intense because it exists within a web of disabilities. And if one problem is solved, and the others left constant, there is little gain." Joyce added a postscript from a later visit to the home, describing the family's further progress. Her last sentence echoes Harrington's notion of a web of disabilities, but with a gentle rebuke to its pessimism. "For a family about to be broken . . . the G. family has taught us that positive strengths may well lie dormant in young families even when they have multiple problems."[20]

Neither Sister Mary Theodora Weick nor Sister Michael Joyce went on to national prominence or scholarly renown. Still, they represent a cadre of health professionals who *saw past the rhetoric of conquest* because their work brought them face to face with the reality of persistent vitamin D deficiency, and who wrestled with its entanglement in the "multiple problems" of America's poor. In the 1960s their work was buoyed by both the optimism of the New Deal and the Great Society and the outrage found in studies like Harrington's. America's growing environmental movement played its part as well, bringing the return of an earlier generation's emphasis on dark, smoke-choked cities. A 1970 feature in *Scientific American* called rickets "the earliest air-pollution disease."[21]

If we take seriously the idea of an Other America of rickets, a foreign land embordered by the cracks in "the country as a whole" where rickets was indeed vanquished, it becomes simple to square the faint thrum of perpetual rediscovery of the "invisible" weight of rickets with the dominant narrative of modern medical progress that saw rickets as little more than a "medical curiosity." This Other America was not a domain defined strictly by race: rickets weighed heavily upon African American children in both urban and rural settings, but no more so than among poor rural white children in many areas.[22] The children in those studies and newspaper articles were not invisible. But they were portrayed as pitiable and vaguely foreign, like the "starving children of Europe," a tone that comes through in Sister Michael Joyce's lighthearted description of steering Mrs. G away from the "Southern style recipes" she was accustomed to. "Yes, pigtails were tasty," she chided, "but they did not have much meat; therefore very little protein to help build body tissue."[23]

What was truly invisible was the distortion of rickety children's pelvises, a "late effect" that could persist long after time had straightened the curved legs of a rickety child. That Gilded Age doctor from Louisville who praised the limb-straightening power of "pot-liquor and fat meat" did not account for, if he even considered, the lasting damage left uncorrected and invisible. Those pelvises, never corrected and now firmly ossified, were for young girls a time bomb set to revisit them years later, on the birthing table. The doctors who studied rickets in the age of vitamin D were well aware of this deadly connection. In an otherwise laudatory editorial from 1970, Harrison noted "the hidden deformity of the pelvis which in the female was a major cause . . . almost certainly of maternal and infant mortality and morbidity."[24] Even if full implementation of universal vitamin D fortification and liberal dosing with cod liver oil *had* led to the disappearance of childhood rickets by the 1940s, the obstetrical aftereffects for yesterday's rickety girls would inevitably haunt the next generation. The middle years of the twentieth century, then, could have been a golden age for studying the link between childhood deficiency diseases and obstetrical complications. The rapid growth in the number and size of American hospitals by then, the skyrocketing rates of hospital delivery, and the strengthening ties between hospitals and medical research institutions would seem to have guaranteed such a project.

And indeed hospitals provided "clinical material" and institutional support for ambitious research into safer (or at least more "efficient") birthing practices, though this research emphasized the ability to exert scientific control over the process of childbirth, with forceps, scopolamine, oxytocin (Pitocin), x-rays to measure pelvic geometry, and the constant presence of a nearby operating room for an emergency C-section should things go south. These researchers did not ignore the role of contracted pelvises; rickets frequently got a shout-out in published studies. But by the mid-century, the bottom line was that rickets was all but cured, and the modern hospital delivery room was well equipped to deal with the present dangers arising from this fast-fading childhood crippler. In a modern obstetrics ward, contracted pelvises should no longer produce protracted labor, a stark contrast to conditions a century earlier, when a pelvis flattened by childhood rickets posed a grave danger to both mother and fetus. The role of rickets in producing the contracted pelvises that sent doctors diving for the forceps can hardly be questioned. The explosion of rickets in seventeenth-century Europe coincided with the production of the tools—in both the theoretical and material senses of the word—that gave rise to modern physician-driven obstetrics.

: :

We don't know which Chamberlen brother actually invented the obstetrical forceps, Peter "the elder" or Peter "the younger," or when those trailblazing seventeenth-century "male midwives" first applied them in their London practice. Apparently, though, it was quite the sight when the brothers arrived to deliver a woman in the throes of prolonged obstructed labor. It took two men to carry the gilded box from their carriage and into the home—the box that onlookers must have assumed contained the heavy machinery that was the secret of the Chamberlens' trade. All but the woman to be delivered were banished from the lying-in room, and even she was blindfolded. As one brother operated the actual "machinery"—the obstetrical forceps the brothers had invented— the other rang bells and produced noises intended to mystify, if not terrify, both the laboring woman and her awaiting family and friends.[25]

The Chamberlen family kept their forceps a trade secret for over a century, but once others learned of it, their instrument established the basic design followed for centuries. At the same time, the Chamberlen family's professional practices blurred, or transgressed completely, the lines between surgeon, physician, and midwife. Their story exemplifies in three generations the path obstetrics would follow in the coming centuries, as male doctors insinuated themselves into the traditionally women-focused profession of midwifery. Peter the elder and his younger brother Peter were both members of the Barber Surgeons Company but achieved great renown in midwifery. Long before they perfected their forceps, Peter the elder served as surgeon and accoucheur to Queen Anne; Peter the younger's son became a physician, thus coming to be known to history as "Dr. Peter." But Dr. Peter also continued in his father's dual role as both midwife and surgeon, at least in that he employed "instruments of iron," as a complaint against him from London's midwives put it. Physician, midwife, and surgeon—the very model of a modern MD obstetrician.[26]

The introduction of forceps and other mechanical devices legitimated the incursions of male medical professionals, whether "man midwife," barber-surgeon, or physician, into the realm of the professional female midwife. Up to then, the male professional's expected role in the birthing chamber was to observe, stepping in only when hope for a successful natural birth was lost. In a world increasingly saddled with the aftereffects of rickets, the new technologies transformed the medical professional's expectations in the face of deadly complicated deliveries. As one historian concluded, only with the new instruments and the publication of new

obstetrical methods in the eighteenth century "did the imagina-
tive horizon of male obstetric practice shift beyond the delivery
of a dead child."[27]

The case for linking this crucial change in professional roles to
rickets rests on two firmly established facts: the rising incidence
of rickets in the seventeenth century and the powerful effect of
childhood rickets in producing pelvic deformities. The combination produced a prolonged crisis of infant and maternal mortality, as well as two centuries of concentrated research into normal
and pathological processes in childbirth. As we saw in chapter 1,
Glisson and the other seventeenth-century physicians can be excused for asserting, sincerely or not, that rickets was "absolutely
a new Disease" in their time. Its rapid ascent toward the epidemic
proportions it would assume must have seemed to be coming from
square one. That leaves rickets' impact on birth complications,
which rests on an older process, the long evolutionary history of
the modern pelvis.

There is a fundamental variable in determining whether a
birth can proceed at all: the difference in the size of the fetal head
and the shortest span in the pelvic opening—the cephalopelvic
proportion. In most mammal species, including some of our clos-
est cousins, there is comparatively ample room for the baby's
head to pass through the pelvic opening. But two of the defining
characteristics of modern humans—we walk erect and have big
brains—were at cross purposes in the evolution of the pelvic floor,
conspiring to create a tighter squeeze, making childbirth in humans more difficult, more dangerous, or—should the proportion
approach zero or a negative value—impossible.[28] The modern human pelvis needs just a gentle nudge during its growth to become
an impenetrable barrier to birth, a nudge that childhood rickets
handily provided.

In the century after Whistler, Glisson, and the Chamberlens,
professionals in the obstetrical arts witnessed the growing menace that rickets posed, correctly linking the disease to the growing
recourse to forceps delivery and the inevitable recourse to fetal
extraction when skilled hands—or instruments, however ably

applied—failed to deliver a live child. A handful of these doctors studied and theorized and published, producing the first scholarship in what would come to be the science of pelvimetry. Percivall Willughby was a successful physician and midwife who practiced in London and Derby in the middle half of the seventeenth century. His career overlapped by a few years that of Dr Peter Chamberlen, but he served a clientele from "the meaner sort of women." In his manual for midwives, which he never saw published, he observes that the pelvises of women who as children "had the infirmity, called the Rickets" lost "their circular roundness" and became "ovall, through which the child will never bee produced, but by violent force of hand, or by some instrument."[29] Later in the century and into the next, the Dutch obstetrician Hendrik van Deventer spent decades treating rickety children and confronting in his obstetrical practice the late effects of the disease. He was a dedicated instructor and advocate for better training of midwives.[30] His successful orthopedic practice may have led him to focus more than his contemporaries on the structures of the pelvis. In his 1701 book *The New Theory for Obstetricians and Midwives*, van Deventer wrote: "It is absolutely necessary to have thorough knowledge of the pelvis; without this we would just be messing around, either with our brains or with our hands."[31]

But it is William Smellie, the Scottish man-midwife, whose name history most closely associates with the themes of "professionalizing" obstetrics after the seventeenth century. Hagiographic essays frequently identify him as "the father of scientific obstetrics," or—inexplicably—"the father of British midwifery."[32] He began practice in his hometown of Lanark in southern Scotland, but in his early forties moved to London to study obstetrics. After a brief course of study in Paris, he returned to London, where he began teaching midwifery himself, advertising a two-year course of lectures for twenty guineas and assuring potential students that "the Men and Women are taught at different hours." He provided services to poor patients who agreed to having his students observe their deliveries. Although he made important technical contributions—most notably improvements in forceps,

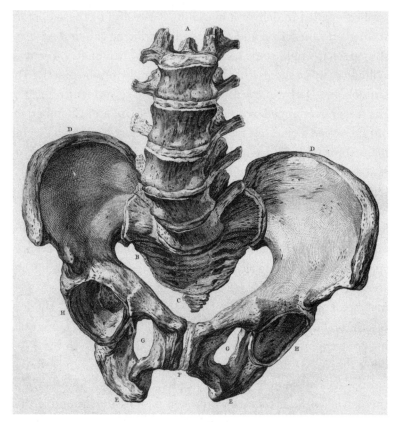

Figure 7.1 "The Third Table, Exhibits a Front-View of a Distorted Pelvis." William Smellie, *A Sett of Anatomical Tables* . . . , 1754. Courtesy of the National Library of Medicine.

encouragement of the study of pelvimetry, and his teaching of the methods for obstructed births—his reputation rests less on his advancement of the science than on his publications: his 1752 *Treatise on the Theory and Practice of Midwifery*, two volumes of his case studies, and most significant, his illustrated obstetric atlas, *A Sett of Anatomical Tables*, published in 1754, containing thirty-nine large plates of vivid anatomical drawings, over half by the Dutch painter Jan van Rymsdyk.[33] The most famous plate depicts obstetrical forceps in use, but the two plates showing normal and rachitic pelvises made explicit the importance of pelvic geometry.

Only a few decades after Glisson and Whistler put rickets on the academic medical curriculum, midwives and physicians developed new birthing techniques and technologies to ameliorate *some* of the fallout from what appeared to be a new disease and began considering more seriously the human pelvis and its role in facilitating or impeding healthy birth. It is not that the earlier doctors' "discovery" inspired the obstetricians; rather the rising incidence of rickets in Europe in the seventeenth century made its presence obvious to both. Rickets may have set the path toward modern obstetrics, but a long road lay ahead. Forceps, anatomical knowledge, and better training could only go so far in averting catastrophe in the birthing chamber. Many a labor prolonged by a flat pelvis, regardless of the outcome for the child, resulted in permanent ruptures in the tissues of the birth canal, a debilitating outcome for which no remedy existed. And a "rickety pelvis" contorted beyond a certain point would refuse passage of the fetal head no matter how skilled the accoucheur, leading to certain death . . . of the child, the mother, or—routinely—both.

: :

Prolonged labor or injury from obstetrical tools can create openings, or fistulas, between the vagina and the bladder (vesicovaginal fistulas) or rectum, resulting in permanent incontinence. While these fistulas are seldom fatal, they can have devastating effects, from physical discomfort to social stigma and psychological struggles. The surgical technique to repair these fistulas, developed in the mid-nineteenth century, marked an early milestone in the development of modern gynecological surgery, and since the "rickety pelvis" was a significant contributor to both prolonged labor and the recourse to forceps, we may credit rickets, if that is the correct word, with a crucial role in that development. Today, vesicovaginal fistula repair is one of the most important gynecological surgeries in parts of the world without adequate medical care. The World Health Organization estimates that over two million women in Asia and sub-Saharan Africa suffer with untreated obstetric fistulas, with between fifty and one hundred thousand new cases every year.[34]

Students of American medical history almost inevitably associate fistulas and their repair with slavery in the Old South, and specifically J. Marion Sims, the southern surgeon who developed the first reliable fistula repair operation, through a four-year program of experimental surgeries on enslaved women. Obstetrics was not a major part of Sims's practice initially; in fact he famously (if facetiously) admitted in his autobiography, "If there was anything I hated, it was investigating the organs of the female pelvis."[35] He learned about fistulas after a local doctor called him in to assist with a difficult delivery. Anarcha, a seventeen-year-old enslaved worker on a neighboring plantation, had been in labor for three days. Sims performed a forceps delivery. Several days later Anarcha developed vaginal openings to both her bladder and rectum. Sims became obsessed with developing a surgical repair. He set up a small "hospital" on his property in Montgomery where he performed experimental surgeries on nine or ten women: Anarcha; two women whom local doctors brought to him (the only others whose names, Betsey and Lucy, Sims recorded); and "six or seven" more he found when he "ransacked the country for cases."[36] In May of 1849, Sims announced he had successfully closed fistulas in Anarcha, Betsey, and Lucy. It was almost four years after his first attempted repair (on Lucy), and after dozens of excruciating surgeries, including over thirty operations on Anarcha alone, all conducted without anesthesia (though with copious opium between operations).[37]

Sims did not report many details about the women who endured his experimental operations, not their names or ages, not even—beyond the three whose names he did record—whether he succeeded in repairing their injuries. We don't know the barest details of their labor, the outcomes, or the presumed causes of their fistulas, beyond the brief description of his first encounter with Anarcha in which he tells us that the "child's head was so impacted in the pelvis that the labor-pains had almost entirely ceased." This of course raises the question most pertinent to this chapter: Did

Anarcha, Betsey, or any of the others have pelvises flattened by childhood rickets? It seems a safe assumption, and many writers examining Sims's experiments have come to that conclusion.

Several years later, Sims moved to New York City, not for glory, his autobiography assured, but to restore his health. After recovering from a prolonged and near-fatal case of dysentery, he set out to build a hospital where he could perform the operation that was already bringing him acclaim. Although he was initially met with cold shoulders by the Manhattan medical world, he rallied philanthropic support for an institution for fistula surgery, and in May of 1855 opened the forty-four-bed Woman's Hospital of New York. The hospital's first patient was Mary Smith, an immigrant from Ireland who had been living for years with a badly botched fistula repair. By now, Sims had become quite proficient at the operation and had demonstrated it to several New York surgeons. But Mary's case was particularly complicated, with masses of scar tissue that had enveloped a bit of wood her Irish surgeon had sewn into place in an effort to block the opening. With echoes of Anarcha's case, it took Sims over a dozen operations, again, without anesthesia, before he was able to close the fistula (although Mary's urethra had also been damaged, so she remained incontinent).[38]

Despite this inauspicious first case, the hospital was a grand success. In the first years, Sims performed hundreds of fistula repairs, sometimes before dozens of observers. Within ten years the institution was outgrowing its forty-four beds and opened a new hospital at Fiftieth Street, on the site of a potter's field established for victims of the 1832 cholera epidemic, its thousands of souls hastily evicted for the new hospital.[39] Most of the hospital's patients were poor, drawing heavily from New York's burgeoning population of Irish immigrants. The hospital's state charter required that at least twenty-five beds be provided free of charge for charity cases. But Sims and the doctors he trained repaired fistulas in women of all classes. The hospital's new four-floor building featured "an elevator of the largest size" for patients' care and comfort, though patients "of the wealthier classes" had no need of the newfangled conveyance, as their $15.50 per week bought them a private room on the first floor. The fourth floor held "the free ward" and outpatient services, with the floors between for patients who paid between $6.50 and $8.50 per week.[40]

Case records for the thousands of fistula repairs performed at the Woman's Hospital could have provided strong evidence tying fistulas to contracted pelvises and hence—even if by inference—to rickets. But while Sims kept reasonably good surgical records, he seemed indifferent to finding either the causes of fistulas or means to prevent them. His successor at the Woman's Hospital, Thomas Emmet, prided himself on his detailed patient records, but historian Deborah Kuhn McGregor found that while Emmet analyzed the statistics "to find the critical factors at the onset of vesico-vaginal fistulas," he "does not include rickets or a contracted pelvis in his exploration of the etiology of the disease."[41] In the 1880s, obstetrician William T. Lusk looked at the Woman's Hospital records and was willing make the inference, concluding "it is impossible to study the cases of vesico-vaginal fistulae reported by Dr. T.A. Emmet without arriving at the conclusion that the existence of contracted pelves is frequently overlooked." Still, McGregor points out, Lusk does not make the final leap, even speculatively, to rickets, despite the ample evidence that childhood rickets was prevalent in the medical histories of the Irish immigrants occupying so many of the hospital's fourth-floor beds.[42]

Fistula repair was the primary surgery performed at the Woman's Hospital in its first years. But Sims and his staff expanded the services in keeping with the broader ambitions of late nineteenth-century gynecological surgery, adding ovariectomy, hysterectomy, clitorectomy, and other "cures" for "women's complaints," from neurasthenia to nymphomania and hysteria.[43] By the early twentieth century, the number of fistula repairs performed in the United States and other high-income nations dwindled; moreover, most of the fistulas needing repair were the result of surgical accidents rather than prolonged labor. By the end of the twentieth century, it was a given that fistulas were "most commonly iatrogenic, due to injury to the bladder during gynecologic and obstetric surgical procedures."[44] By no means should this suggest that rickets no longer flattened young women's pelvises with potentially grave risk to childbirth. Instead, the disappearance of obstetric fistulas was due to other tools available to midwives and obstetricians when a constricted pelvis threatened mother or fetus, tools including more accurate measurement of pelvic geometry, interventions in the natural course of pregnancy, and what was once the last recourse—delivery by cesarean section.

: :

The science and art of measuring the human pelvis had come a long way since the 1700s, when van Deventer cautioned that ignoring it would leave obstetricians and midwives "just . . . messing around."[45] As the business of childbirth grew more enmeshed in biotechnical medicine, accurate measurement could powerfully shape the course of labor and delivery. Of course, a normal finding provided both the mother and her caregivers some assurance of an uncomplicated birth. On the other hand, the finding of an obstructed pelvis was equally instructive. When determined early in the pregnancy, the pregnant woman and her caregivers had options to prepare for an easier labor, including severely restricting the mother's diet in the hope of producing a smaller baby or inducing labor before the fetus reached its full size.[46] If the obstructed pelvis was detected late in the pregnancy, the options narrowed: a spontaneous labor might be tried, but with instruments at the ready. Increasingly over the course of the twentieth century, the immediate recourse for a seriously obstructed pelvis was a cesarean section.

The system most widely used to classify the various pelvic shapes defines four, based on shape. One of these, the platypelloid, or "flat," pelvis is oval but wider side to side. The "rachitic pelvis" is a form of flat pelvis.[47] But it is important to note that while "rachitic pelvis" is the only classification with rickets in its name, studies in the early twentieth century implicated vitamin D deficiency disease in several of the other types of constricted pelvis.[48] By the late nineteenth century, well-trained obstetricians understood the value of pelvimetry, and they and their hospitals collected data that linked rickets and birthing complications. For example, in 1899 Baltimore obstetrician J. Whitridge Williams analyzed the data on contracted pelvises in the first thousand deliveries at Johns Hopkins Hospital. Overall, he found that 13 percent of the thousand women had obstructed pelvises, but the rate for the 469 Black patients in the study was 20 percent, as compared with 7 percent of the 531 white women.[49] That is fully consistent with typical reports on the incidence of rickets across racial groups in American cities of the day. Edwin Bradford Cragin's 1916 obstetrics textbook reported a rate of constricted pelvises of about 9 percent in twenty thousand consecutive cases at New York's Sloane Hospital for Women. He described rickets as "the disease which is most often the cause of pelvic deformity in this country, especially among the colored race."[50]

When these doctors compare various forms of pelvic constrictions across racial lines, especially as they relate to rickets, it becomes tempting to think back to those lighthearted news reports from the late nineteenth century joking about what a common sight it was to see all those bowlegged white beachgoers and ballerinas, and wonder whether obstetricians were juking the stats to protect the sensibilities of their white patients or to fit the data to their expectations about race and rickets. Cragin found marked differences in the rates for different types of distorted pelvises in his "black" and "white" patients. Sloane Hospital classified entirely by shape, not cause, and so did not include "rachitic pelvis" as a category. Still, 75 percent of white patients were recorded as having a "simple flat" pelvis, compared with 43 percent in the Black group. One third of the Black patients' pelvises were of the "generally contracted" variety, nearly three times the proportion for white women. Williams's 1899 analysis of a thousand cases at Hopkins found that that 6 percent of the deformed pelvises among white women were from rickets, compared with 22 percent for Black women.[51]

Williams revisited his statistics two years later and concluded that "every fourteenth white and every sixth colored woman in

Baltimore has a contracted pelvis." The "simple flat" was the contraction found most frequently in white women, and "rachitic" the least. For Black women the most common shape was "simple contracted," followed by "rachitic," and finally "simple flat." His seeming precision leads to the question of how subjective the distinction between "flat" and "rachitic" (a specific form of flat pelvis) might have been. Williams was surprised that rachitic pelvises were not more common in the Black patients; after all, anyone observing "the colored people in the large cities of the South cannot fail to be impressed with the marked prevalence of rachitis among them." It seems that in 1901 Williams did not yet appreciate how frequently rickets could produce a "generally contracted" pelvis (circular in shape but small)—a fact he recognized in a later article. He explains away his surprise by falling back on the popular geographic determinism of his time that assumed African Americans belonged in the fields of the tropics, concluding that "colored people in the large cities are in great part physical degenerates . . . and this manifests itself in the pelvis as well as in other portions of the body."[52]

Williams had reexamined his data for an article encouraging his colleagues to stand ready to perform a cesarean section when confronted with an obstructed pelvis. At this point, Hopkins had performed only three cesareans, with a maternal mortality of 33 percent.[53] But statistics on cesareans performed elsewhere convinced him that maternal mortality for the operation was less than 4 percent when performed early in labor. And so, he argued, "its employment should be markedly broadened."[54]

Twenty years later he was even more enthusiastic. By then Hopkins had performed 183 deliveries by cesarean, with a "gross mortality" of 5.46 percent. At that point he concluded "that the operation is being abused throughout the country." Not that doctors were performing too many cesarean sections. The "abuse" he warned of was their *reluctance* to perform it except as a last resort. Letting a hopeless delivery sabotaged by an obstructed pelvis grind on through repeated manipulations and fruitless attempts at instrument delivery, he argued, leads to an exhausted mother with stressed, abraded, and potentially infected tissues, ingredients for a much riskier operation. Better attention to pelvic geometry would have shown the need for a cesarean early in labor or even before, the operation performed under controlled conditions with minimal risk.[55]

We can see in Williams's admonition the dynamic that would lead to the situation today, when close to one-third of American babies are delivered by cesarean section, the most common major surgical procedure in American hospitals.[56] Williams's stance is actually quite conservative: cesarean section is the correct choice only when it is the safest alternative. But as more are done, safety improves, changing the equation in favor of still more abdominal deliveries. It was not a gradual transformation, but a long, slow rise followed by an eruption.

At the opening of the twentieth century, cesareans were extremely rare. By 1920 Johns Hopkins had delivered approximately 20,000 babies, only 183 (0.9 percent) by cesarean. At midcentury the nationwide rate was still below 3 percent. But between 1965 and 1990, cesareans skyrocketed to over 25 percent of all births. Historian Jacqueline Wolf identifies shifting risks and—just as significant—the *perception* of those risks as playing the defining role of in this sudden inflection: "While before the 1970s doctors and their patients almost always saw the risk as being inherent in the surgery itself, in the 1980s and after, the risk seemed to be in not performing the surgery."[57] While that is essentially Williams's argument in 1921, his enthusiasm for cesareans was reserved for cases where disproportion of head and pelvis made other options riskier, as was the case in 78 percent of Hopkins cesareans up to that time. He wanted to see more cesareans, but only those that lowered risks to the mother. And the best way to lower mortality "in case of disproportion" was "learning to determine before the onset of labor whether operation will be required or not"—safer surgery through accurate pelvimetry.[58]

When cesareans were still high risk, practical knowledge of pelvimetry and advanced technique in vaginal delivery were essential to risk reduction, but as cesarean section became safer over the century and the risk calculation shifted in its favor, the obstetrician's need for deep knowledge of pelvimetry faded. In 1903 Williams had devoted seven chapters of his new obstetrics textbook to the pelvis, including ten pages of practical instruction on pelvimetry. And a quarter century later, with cesarean mortality figures still hovering around 5 percent, Williams insisted that "any one with two hands and a few instruments can do a cesarean section, but . . . it frequently requires great intelligence not to do it."[59] Imparting that intelligence was central in Williams's *Obstetrics*, which continued to devote chapters to the pelvis through its fifth edition of 1923. But gradually the pages on the pelvis gave way to new concerns. In the 1941 edition, the chapter titled "The Pelvis," which had appeared on page one of the first edition, disappeared, "The Female Organs of Generation" taking its place of pride; by century's end all of the other chapters on the pelvis disappeared as well, with the remaining related content making up less than 1 percent of the book's pages.[60]

Safer cesareans may have reduced the need for deep training in pelvimetry, but other forces pushed pelvimetry from center stage as well. Perhaps most significant was the trend toward replacing the obstetrician's pelvimeter with the radiologist's x-ray machine. Radiology's potential for revealing hidden flaws in the pelvis was obvious from the earliest years after Röntgen's discovery in 1895. By 1940 radiopelvimetry had become, in the words of Yale obstetrician Herbert Thomas, "a routine prenatal procedure," a status it maintained into the 1970s, when concerns over radiation exposures overtook enthusiasm for its diagnostic powers.[61] The perceived superiority of radiopelvimetry's precision helped amplify a growing frustration many physicians expressed about the imprecision and subjectivity of both external and manual (internal) pelvimetry. The pelvimeter, a device obstetricians had relied on since the eighteenth century, came under especially harsh judgment.[62]

The final and obvious reason pelvimetry came to be less crucial in obstetrical decision-making brings us back to the theme of this chapter: the dramatic reduction in the incidence and severity of vitamin D deficiency in the mid-twentieth century. The babies who got cod liver oil in the 1920s were less likely to have "rickety pelves" in their twenties. Consider two large hospital surveys from Baltimore and Washington, DC, from before and after the 1920s. "Disproportion of the pelvis" was the indication listed for 79 percent of the 183 cesareans performed at Johns Hopkins before 1921, with three-fourths of those presumed to be from rickets. Twenty years later, Freedman's Hospital in Washington attributed 47 percent of their cesareans to "Cephalo-pelvic disproportion," and from 1959 to 1969 that rate dropped to 29 percent. While the actual rate is confounded by how second and subsequent cesareans were counted, the authors of the study were clear: obstetricians had new conditions to worry about, but "tuberculosis, rickets, eclampsia and heart disease are rapidly fading as indications for cesarean section today."[63]

For Williams and others at the opening of the twentieth century, the future of safer cesareans and pelvimetry were intertwined, and childhood rickets was a major driving force for both. Williams saw this connection as particularly true for African American women and speculated that if "suitable dietetic and hygienic measures should eventually lead to the disappearance of rickets, Caesarean section would be very rarely indicated in the black race."[64] While the "dietetic measures" of vitamin fortification no doubt reduced the number of dangerously constricted pelvises in both Black and white women, the march toward today's reliance on cesarean delivery continued apace. Many young girls acquired rickets in the age of fortified milk, in the years after "the disappearance of rickets." Eighty years after Williams's prediction, they still did. Many of these women ended up getting cesareans. But not for "contracted pelvis" or even "cephalo-pelvic disproportion." Now they made up an undifferentiated portion of the millions delivering by cesarean each year. A patient's history of rickets was of no consequence. Birth by cesarean section had become another technological fix for vitamin D deficiency, or another way to disregard it.

: :

To paraphrase the sunny rhetorical question from the 1950s that opened this chapter: Has vitamin D milk accomplished what J. Whitridge Williams hoped it would? While by most measures, vitamin D deficiency is a shadow of its former menace and rickets-related pelvic contractions are rare, rates of cesareans continued to climb anyway, and the higher rates for Black women persisted, 13 percent higher than average in 2019. Maternal mortality disparities are even more pronounced, with African American women over two and a half times as likely as white women to die from pregnancy and childbirth.[65] There are many reasons for these disparities, and the near disappearance of rickets-related pelvic contractions doesn't let vitamin D deficiency off the hook. A provocatively titled 2009 overview of "Vitamin D and Pregnancy: An Old Problem Revisited," ran through a partial list: "Vitamin D deficiency during pregnancy has been linked with a number of maternal problems including infertility, preeclampsia, gestational diabetes and an increased rate of caesarean section." Like so many findings related to nutrition and vitamins, the evidence for the links are mixed and often contradictory. As for solutions, it is not a simple matter of assuring everyone gets their "D," as the authors cautioned: "The optimal concentration of 25 hydroxyvitamin D is unknown and compounded by difficulties in defining the normal range."[66]

Difficulties indeed.

No Magic Bolus

Rickets, Risk, and Race in the Age of Uncertainty

From 2002 to 2003, a sensationalized story featuring an infant with rickets burned through New York's tabloids. Officials from New York City's Administration of Children's Services had taken fifteen-month-old IIce (pronounced "Ice") Swinton into protective custody in November 2001. The tiny child weighed only ten pounds, had no teeth, was suffering from frail and broken bones, and exhibited the distinctive signs of rickets.[1] In April of 2002, Queens County prosecutors charged the infant's parents with reckless endangerment, a charge soon amended to include first-degree assault. Details of the "Vegan Baby Trial" peppered local and national papers for months. *New York Post* and *Newsday* readers learned of the parents' dedication to veganism, how IIce was fed a diet of ground nuts, beans, juice, and herbal tea, but neither breast milk nor baby formula; they read of child services officers who failed to act to defend the infant; and, in May of 2003, of the trial's outcome, with both parents sentenced to at least five years in prison.[2]

The lurid reporting of baby Swinton stands in stark contrast to the stream of clinical studies percolating through public health and medical journals at the time warning of an increase in clinical rickets, sounding the alarm of the "reemergence of a once-conquered disease."[3] Most of these cases, like the Swinton baby's, involved African American children whose parents swore off both commercial infant formula and prescription vitamin D drops. Typically, the mothers had chosen to breastfeed their babies (though this was *not* the case with the Swinton baby). What distinguished the reporting on the Swintons was its sensationalism and blanket condemnation of the parents, in language that placed the Swintons in a "culture of poverty." The parents' veganism came in for equal condemnation as well, under headlines that ranged from "Tragic Tot's 'Rotten' Diet," "Was Tot's Vegan Diet Criminal?" to "Prison for Vegans."[4]

But the comparison of greatest relevance to this chapter is how issues of race figured in the two sets of reports—the clinical and the sensational. What counted most in the Swinton case was not race but the parents' behavior, especially their dietary practices. Newspaper reports only emphasized their racial identity when the story turned to local Black advocacy groups criticizing how the criminal justice system and the press treated the Swintons, and then protesting the verdict and sentencing after a mostly white jury convicted the couple.[5] And while the scientific reports also discussed dietary and behavioral choices, race featured prominently in their discussions, no matter how circumspect the language. An editorial in a pediatrics journal described a new "wave" of rickets in the United States, "found among dark-skinned infants, particularly among the African American children, whose mothers are breast feeding." A report on a rickety infant in Atlanta skirted naming racial categories but still made hard-and-fast pronouncements based on skin color: "Individuals with darkly pigmented skin require more sun exposure to synthesize vitamin D, thus, placing this dark-skinned patient at higher risk." Another report highlighted five new cases, all "breastfed black male children who did not receive vitamin D supplements while breastfeeding."[6] Perhaps the difference in focus was due simply to the minor role rickets played in baby Swinton's health problems; for the daily press, theories about race and rickets simply were not pertinent to the story they were telling of gross neglect and willful disregard of childrearing norms.

Whether found in newspapers or medical journals, these reports address questions at the heart of this book—questions about environment, economic status, dietary preferences (especially infant feeding), racial identity, and the biological role skin color plays in rickets. These questions press me to address more explicitly an argument that has been insinuating itself throughout the first seven chapters—that rickets, entangled as it is in environmental, cultural, and socioeconomic factors, exemplifies how diseases can be both biologically and socially constructed. Rickets makes an especially tricky case study, in large part because of the complexity of its etiology ("It's environment! No, it's diet!"), but also because of the presumably clear role skin pigmentation plays in the UV-driven photosynthesis of vitamin D. Complicating these questions even further, this chapter takes more seriously the broad spectrum of vitamin D deficiency within which rickets is merely the extreme expression. Considering vitamin D deficiency and its potential role in health—whether in infants, children, young adults, or the elderly, and across ethnic and racial categories—increases the complexities exponentially. But instead of even pretending to untangle and elucidate all of the complexities, this chapter skims across a very wide, very deep pond, beginning with the earliest days of humankind and the evolution of the thousand natural hues that flesh is heir to, coming at last to rest in a modern clinic examination room. It is here where health professionals, charged with advising and prescribing for their patients' best interests, must navigate the narrow straits between acting on or ignoring those "long established" associations between vitamin D deficiency and racially defined populations, or, perhaps, to chart a different course entirely.

: :

Two stories of migration drive the ancient and modern history of rickets; both are intertwined with stories of sun, skin, and bones. First comes the eons-long migrations that shaped human evolution, beginning with the first humans to spread out over Africa and across the continents, eventually making their homes nearly everywhere, from tropics to tundra, with dietary and cultural practices as varied as their environments, environments that spanned vast differences in the quantity and intensity of sunlight. Many thousands of years later came the migrations told of in history books, with people from long-settled populations moving, or being moved, from their ancestral environments to dramatically different new ones. Compared to the movements of prehistoric peoples, the modern migrations were instantaneous, whether they covered great distances to new continents and climates, or shorter, but equally jarring, journeys from open countryside to cramped and sunless cities. The ancient migrations produced the dazzling array of population variations that define humanity, including a wide range of skin color and "normal" vitamin D metabolisms. The modern migrations loosed those bodies from their accustomed environmental and cultural moorings or threw together people from long-isolated continents. The resulting efflorescence of rickets in recent centuries is just one among many health consequences of the rapid movement of people, and skin color seemed to figure deeply in explaining its rise.[7]

It is common for brief surveys of the evolutionary origins of skin color to focus on the origin of "white" skin, a tale often told as the inevitable response to selective pressures of a dark-skinned species moving from the UV-drenched tropics to high latitudes, where sunlight is less intense. This origin story is often framed as a move from the "torrid heat" of the tropics to the normative "temperate regions." This Eurocentric narrative can drift into whiggish (and racist) celebration of plucky Caucasians whose very bodies put in the hard work of adapting to the harsh realities of the promised land above the fortieth parallel—Euro centered and vitamin D centered.

The true starting point is much further back, at least another million years, with the evolution of *dark* skin. If chimpanzees, our modern fur-covered primate cousins are any guide, our hairy hominid ancestors had pale skin beneath their fur. Shedding that fur required that these "naked apes" find a new way to shield themselves from the sun, and melanin fit the bill. The evolutionary balancing act between UV light and skin color involves the sun's effects on two vitamins—vitamin D and folate. Sunlight on skin produces the one but can destroy the other. Vitamin D probably had little to do in the transition from light skin to dark. This may be surprising if we assume that melanin protected the sun-baked body from producing toxic levels of vitamin D. But the skin's vitamin D metabolism is self-regulating: in the event of excess UV exposure, the active precursor molecule (pre-vitamin D3) breaks down into inactive molecules rather than being converted to D3 and passing to the bloodstream.[8]

The effect of UV radiation on folate is a different matter. Folate, a type of vitamin B9, is found in many foods from beans to beef liver. It is essential in fundamental metabolic processes such as making blood cells, creating and repairing nucleic acids (DNA and RNA), and amino acid metabolism. Folate deficiency is linked to birth defects of the nervous system, including spina bifida and anencephaly. So clear is this link that folate is a critical ingredient in prenatal vitamins, and over fifty countries mandate wheat flower to be fortified with folic acid (a synthetic form of folate). The reproductive hazards of folate deficiency would have been deadly to fair-skinned early humans and likely provided powerful selective advantages to those with a trait for melanin-rich skin.[9] Dark skin may have been helpful in other ways, such as providing a more powerful barrier to microbial, parasitic, and waterborne diseases. The reviews are mixed on the role skin cancer played in selecting for dark skin. Most fatal skin cancers develop late enough in life as not to interfere with reproduction, so protection against skin cancer would offer little selective advantage for dark skin. But some studies of cancer and albinism suggest that fatal cancer could develop before reproductive age in individuals with very low melanin levels, posing a potentially powerful selective advantage. None of these factors, though, were as powerful as the need to protect folate.[10]

The migrations out of Africa began approximately sixty thousand years ago.[11] As humans moved further from the equator, to where the sun rode lower in the sky and seasonal fluctuations in sunlight hours deepened, the importance of balancing vitamin D production and folate protection became clear. The effectiveness of dark skin in filtering UV became a problem, as vitamin D deficiency would certainly have plagued them. The need for vitamin D initiated the second phase in the evolution of modern human skin colors, selective depigmentation. The many ways vitamin D deficiency provided selective advantage to fairer skin are not limited to rickets or other skeletal effects. Low vitamin D is implicated in an array of reproductive hazards: for the mother, preeclampsia and cardiac problems resulting from calcium metabolism disruptions; and for the newborn, low birth weight, tetany, and another suite of potential calcium disorders.[12]

Turning to rickets proper, clear and powerful selective factors for pale skin came into play. While severe rickets *can* be fatal, its impact on infant mortality was unlikely to have been a driver for depigmentation. Pelvic deformities, on the other hand, are a common result of severe rickets and could have applied great selective advantage for light skin, as colorfully suggested by an American anthropologist in 1934: "Rickets is not a deadly disease to the individual even in the severe form that is deadly to the race by inhibiting reproduction."[13] Rickets probably had little impact on the evolution of characteristic pelvic shapes for people of different climates. Long-established populations may have average pelvic shapes that differ markedly from other populations across the globe, but vitamin D had little to do with it. Instead, a characteristic pelvis shape tends to align with other shared traits in a population whose ancestors passed through an evolutionary "bottleneck," say when a small migratory population with relatively low genetic diversity evolves in isolation from the groups they left behind, as with the Paleolithic hunter-gatherers who migrated over the Bering land bridge into the Americas twenty thousand years ago.[14]

One more adaptation in skin pigmentation was necessary. As humans moved further from the equator and their baseline melanin levels dropped, it became important for individuals to be able to adjust melanin density with the changing seasons. Tanning was a critical mechanism wherever deep seasonal fluctuations in UV made a one-shade solution problematic. Untanned skin in the winter sopped up the limited supply of UV light for vitamin D production but threatened folate retention when summer came. A mechanism that gradually added melanin to the skin through spring and withdrew it as winter approached assured both a steady supply of vitamin D and protected the body's folate stores. The genetic instruction to "tan, don't burn" also protected the fair-skinned from painful and potentially dangerous sunburn.[15]

Things got more complicated in the last twenty thousand years. For one thing, migrations were always a process, not a path to a single destination, even when geographical isolation held populations in one place (the disappearance of the Bering land bridge for example). As global migration continued, the complexity of populations multiplied, with returns, divisions and reunions, and conquests and retreats bringing genetic admixture at every turn. With the first agricultural revolutions of about ten thousand years ago, cultural practices and technology became increasingly important—often driving humans' ongoing restlessness and liberating settled populations from the strict algebra of the latitude-pigmentation equation. Clothing made forays into the coldest regions possible and made the semitropical sun survivable for fairer-skinned peoples. It is not that genetic adaptations ceased with technological change. On the contrary, sometimes technologies elicited lasting genetic responses.

The rise of agriculture and animal husbandry introduced new foods humans had to learn to digest and hosts of microbiological threats from the animals they lived with and the soil they tilled.[16] One such adaptation, tangentially related to the history of vitamin D, gave rise to the strong divide between populations' ability to digest milk. Western medicine tends to treat lactose intolerance as a pathology, when it might be more accurate to pathologize what can colorfully but accurately be described as "the aberrant phenomenon of lactose *tolerance*."[17] Lactose is the primary sugar in animal milk; all infants comfortably digest lactose. But for nearly two-thirds of the earth's population, that ability disappears after the first few years of life when their bodies stop producing lactase, the enzyme in the gut that breaks lactose down into the two readily digested sugars glucose and galactose. Starting around ten thousand years ago, herding peoples from a few widely separated regions added liquid milk from their domesticated herds into their diets, and in a flash by normal evolutionary standards, developed what biologists prefer to call lactase persistence. In Europe, the new trait really took off, so that today roughly 80 percent of Europeans drink milk without difficulty. It bears repeating that milk, whether human or bovine, is not a natural source of vitamin D, a fact easily forgotten in the age of "universal" vitamin D fortification, leading some to erroneously assert that one-third of the world's population came to drink cow's milk because their ancestors needed it as a source of vitamin D![18]

::

Transportation, trade, war—the processes that brought the first great cities and emerging empires—also brought the knowledge that humanity was made up of people of many skin colors, launching the long history of theorizing about links between race, geography, and pigmentation. Hippocrates and Galen folded skin color into their theories about how bodies adapt to climate and environment by balancing their four essential fluids, or humors (black bile, yellow bile, phlegm, and blood). Carl Linnaeus, the eighteenth-century Swedish botanist, formalized the association between humor and skin color, classifying humans into four races: *Homo Europaeus albescens*, *Homo Americanus rubescens*, *Homo Asiaticus fuscus*, and *Homo Africanus niger*; later that century, Immanuel Kant theorized that all four races, defined by skin color, had emerged from one European "stem genus: white brunette." Darwin took things in a different direction. He belittled simplistic classifications based on skin colors and placed Europeans not at the origin but the apex of evolution.[19] The twentieth century arrived fully armed with its armada of scientific racism, Darwinism (of various stripes from social to Christian), and twin commitment to the promise of eugenics at home and global imperialism abroad. Consequently, the theories of race, skin color, and geography multiplied, from the environmental determinism of Ellsworth Huntington to Leonard Jeffries's theory of "Sun People" and "Ice People."[20]

The modern science of rickets developed in this heady time for racial theorizing, and the explanations for rickets' disproportional impact in people of color naturally reflected the racist proclivities of the era. These conversations can be mapped out over three overlapping phases. Through the early twentieth century, arguments for a particular ethnic group's "racial" susceptibility perhaps surprisingly almost always emphasized social and economic factors, not race. Then, as vitamin D science matured and the myth of "universal" fortification took hold, explanations for stubborn racial and ethnic disparities relied on physiological differences more than differences in social and economic conditions. Since early this century, we may have entered a third phase, with reports of clinical rickets showing greater nuance as to the social and cultural factors shaping racial disparities and the apparent rise in cases.

In the years after the Civil War, some southern physicians made a strong case for what today would be called the social determinants of health for African Americans, though that may not have been their intention. Their arguments were often couched in the harshest racist language, but often spoke to social, not racial factors. They asserted that under slavery, African Americans young and old had been well cared for. They admitted that bowlegged infants were a common sight in the slave quarters but argued that with the food their enslavers provided, those legs grew straight.[21] But, echoing slavery's apologists, they insisted that when the Civil War freed the slaves from their forced servitude, it also removed them from the "generous care" slaveholders had provided. Surely, ran the argument, a "redeemed" South, its economy rising again, would restore the freedman's health.[22] When Baltimore obstetrician J. Whitridge Williams predicted in 1921 that dietary improvements would lead to the end of rickets and the need for most cesarean sections in "the black race," he tapped into this tradition. Even the most blatantly race-focused explanations in the early twentieth century relied on social factors to explain freedom from rickets in populations whose race—if skin color were the only factor to consider—would predispose them to it: "Among savage peoples the children are, in general, free from [rickets], because they are exposed to sunlight every day. The Eskimo baby is free from it, even though he lives in a dark hut, because he is suckled by a mother who consumes great quantities of animal fat and oil."[23]

Ironically, after the discovery of the "magic bullet" of vitamin D and its presumed "universal" presence in the American diet, explanations for population differences in rickets came to rely more directly on race than before, often with an unspoken defeatism: rickets had been conquered except "among the dark skinned races," for whom there was no end in sight. Early in the age of vitamin D, Alfred Hess referred to a "constitutional factor" in Black people, apart from the "well recognized factor" of skin pigmentation; "the negro" he speculated, "may possess an inherent racial tendency to rickets."[24] The new knowledge of vitamin D, especially as its production in the skin was concerned, did not displace old presumptions about race; it hardened them. Rickets in dark-skinned people was no longer a sign of cultural inferiority, no longer "the physical deterioration of a southern race in a northern climate." Now it was all about the melanin and latitude.[25] How other than racial essentialism was one to explain the persistence of rickets in African Americans in light of the "universal" availability of antirachitic food supplements? With better understanding of the mechanisms of UV light, skin pigmentation, and the production of vitamin D, it became common to argue that not culture but melanin determined rickets' racial proclivities. "Dark-skinned races," a 1970 report asserted, "suffer an additional disadvantage, since black skin can filter off up to 95% of the ultraviolet light which reaches it." A more recent report suggested African Americans suffered from a "latitude–skin color mismatch"—that their skin pigmentation "is not appropriate for the solar UV doses at various latitudes."[26] The higher incidence of rickets in African Americans would not be explained by their sun exposure nor the content of their dairy products, but by the color of their skin.

The situation in other parts of the world since the 1960s reinforced the view of persistent rickets as a racial or racially defined cultural problem. In the United Kingdom, immigration from Britain's former colonies in South Asia rose dramatically in the 1960s, accompanied by reports of clinical rickets and osteomalacia among the new arrivals. The medical press promptly rechristened this resurgence of the "English Disease," as "Asian Rickets." In the intervening years, medical professionals across Europe have made similar findings, with rickets commonly appearing among Turkish immigrants in Germany and African and Middle-Eastern immigrants in France; a 2016 survey found the same story of rickets among immigrants from tropical lands in medical reports from Italy to Finland.[27] Observers pointed out cultural factors— dietary habits (South Asian immigrants preferring unfortified ghee over Britain's fortified margarine, for example) or traditional full-length clothing that filtered UV light. But they almost always came back to skin color and the "latitude–skin color mismatch," as one study rather meekly expressed it: "Premigration latitude is an important factor for vitamin D deficiency in immigrants."[28]

Studies of rickets in the lands where those immigrants came from suggest that latitude and skin color explain less than some thought. Rickets, osteomalacia, and severe vitamin D deficiency are intractable problems in parts of India and the Middle East, and while diet plays a role, cultural practices, notably full-body covering for women and other practices of purdah, severely limit UV absorption, producing vitamin D deficiency in both mothers and infants. A study in 1992 found a rate of rickets in Indian Hindus one-third that of Muslims, who are much more likely to practice purdah.[29] In other settings childhood rickets is due almost entirely to diet. Studies in South Africa and Nigeria find that, despite more than adequate sun exposure, children there frequently develop clinical rickets, caused by inadequate calcium intake—a situation reminiscent of the conditions on southern slave labor camps in the United States.[30]

: :

If talk about rickets nowadays tends to focus on race, vitamin D *deficiency* is a very different story: the popular press and scientific literature treat deficiency as a problem for young and old, Black and white. "Deficiency" as it is used today defines a spectrum disorder defined less by clinical symptoms than by vitamin D levels in the blood. Seen this way, clinical rickets is simply "the end stage of vitamin D deficiency."[31] In the 1930s, the term of art was "subclinical rickets," hinting at the modern concept of a continuum. Confusing matters, soon after vitamin D and its causative role were known, clinicians did start referring to "vitamin D deficiency rickets," but this was less to imply a quantified spectrum of deficiency than to distinguish rickets caused by a deficiency of the vitamin from the rarer forms caused by vitamin D resistance or calcium deficiency.[32] At the start, only clinical rickets was a clear target for study and eradication; to the degree researchers thought about a continuum of deficiency, they seemed to assume that those deficiencies would fade away with the coming victory over rickets. It was only in the last third of the twentieth century, with an effective response to rickets apparently underway, that the two streams—rickets prevention on one hand, and on the other, research into the subclinical effects of low vitamin D levels and defining optimal vitamin D levels and how to achieve them— became separate paths of inquiry.

Early in the vitamin D era, pediatrician Edwards A. Park declared: "But for rickets vitamin D would not have been discovered."[33] It seems equally true that but for vitamin D blood tests, vitamin D deficiency would not have been discovered, along with all its controversies and inconsistencies. Until late in the last century, there was no way to study or understand "vitamin D deficiency" as a spectrum in any meaningful—that is to say, quantifiable— way. Methods for local clinical laboratories to do routine assays appeared only in the mid-1980s, launching the search for the ideal numbers: "normal" or "optimal" vitamin D levels and consistent standards defining levels of risk: "deficiency," "insufficiency," and

so on.[34] These standards were established in the service of finding predictable supplementation—all defined by precise measurements but frustratingly uncertain in terms of measurable health
outcomes.[35]

With the ability to quantify individuals' vitamin D levels came
the ability to map their "D" to statistical analyses of enormous data
sets and find associations between a given level and a given health
problem—diagnosis by association.[36] Naturally, the first links to be
tested were the skeletal effects of low vitamin D such as osteoporosis, bone pain, muscle weakness, falls, and low bone mass; some
of these had been routinely treated with cod liver oil long before
the discovery of vitamin D.[37] But by the end of the twentieth century, studies linked vitamin D deficiency to nonskeletal conditions,
including cardiovascular disease, hypertension, diabetes, bacterial
and viral infections, autoimmune diseases, and cancer.[38]

This glut of possible associations between vitamin D and specific body systems and functions is literally built into our DNA.
Because of vitamin D's essential role in calcium metabolism and
calcium's ubiquitous role in all body systems, most human tissues contain the vitamin D receptor—about four hundred different tissues and cell types—and the human genome has at least a
thousand vitamin D targets.[39] Investigators may reasonably hy-
pothesize that any organ or process that relies on calcium will be
affected by a deficiency of vitamin D; if they then find an association through experimentation or statistical analysis of large data
sets, the presence of those genetic targets reinforces, or justifies,
their conclusion. Vitamin D's role in immune function is a clear
case in point. The arguments for why it *should* be critical in fighting
infection are unassailable: first, a century of vitamin D therapies
employed against tuberculosis—both environmental (heliotherapy) and dietary (via cod liver oil); second, modern biochemical evidence that immune cells express vitamin D receptors and
that the immune system produces the active form of vitamin D,
which in turn induces the production of antibacterial products;
and, finally, epidemiologic studies linking vitamin D deficiency
with greater infectious disease burden.[40]

And so, when the COVID-19 pandemic arrived, the fevered search for preventives and cures naturally included a possible role for vitamin D. Large vitamin D studies already underway added the novel coronavirus into their designs; new studies launched, specifically looking for links between deficiency and COVID outcomes, while others measured the impact of high doses on COVID patients. From early in the pandemic, it seemed clear to many observers that COVID's impact fell lighter in sunny regions closer to the equator. Studies that accounted for confounding variables from urbanization to national health expenditures would confirm what seemed obvious.[41] And so, while the president blithely assured that COVID would go away "with the heat—as the heat comes in," the pandemic thundered on, and in the months before reliable vaccines and powerful drugs could turn back the juggernaut, vitamin D played a prominent role in holding actions, both in hospitals and in the hopes of worried citizens.[42] But was the vitamin any help? By spring of 2021, the reviews were starting to come in. They were mixed: a press release for a University of Chicago study declared "Study Suggests High Vitamin D Levels May Protect against COVID-19, Especially for Black People," while a similar release for a McGill study cautioned "Vitamin D May Not Protect Against COVID-19, As Previously Suggested."[43] A shaky consensus rose that being vitamin D deficient increased the risk of contracting a COVID infection and made respiratory symptoms worse, but that dosing people who already had sufficient levels of vitamin D provided little or no added advantage.[44]

The ambivalence about vitamin D's value in a global pandemic reflected the lack of consensus that long characterized vitamin D research. For over two decades, epidemiological and experimental research had continued to find new health benefits from boosting vitamin D levels, often followed by epidemiological and experimental research that refuted those findings. A major dust-up over the issue arose in 2011, when the Institute of Medicine issued a position paper firmly arguing against growing calls to raise the threshold defining vitamin D deficiency from 20 ng/ml to 30 ng/ml.[45] The IOM committee found "no basis to support . . . causal relationships" between higher vitamin D levels and prevention of nonskeletal health issues such as cancer, diabetes, or infectious disease and worried that adopting the 30ng/ml standard would create a low vitamin D pandemic by fiat, leading to an increase in laboratory tests of patients at little risk and more spending on vitamin supplements that added no health benefit, and, when overused, raised the risk of hypercalcemia and other conditions.[46]

Proponents of the higher levels rejected by the IOM responded with confident warnings about the health costs of the status quo. Robert Heaney, a specialist in osteoporosis, and Michael Holick, a leading vitamin D researcher, worried that the IOM recommendations would lead to lower average vitamin D consumption. "While a small fraction of consumers may well have all the vitamin D they need," they argued, "on balance, we consider a general downward trend to be harmful to the health of the public." The IOM's approach to setting safe minimal, optimal, and maximum levels erred in treating vitamin D "as drugs, which are foreign chemicals." The standard for vitamin D levels should not be the status quo in modern sun-starved bodies but should reflect the conditions "that prevailed during the evolution of human physiology," when vitamin D levels would have ranged between 40 and 80 ng/mL. Heany and Holick sniffed at the IOM's fears of vitamin overload, accusing the committee of using far lower standards for papers highlighting risks than for those indicating the benefits of higher vitamin D levels.[47]

The uncertainty extends even to the field of pediatric orthopedics, the home court for studying vitamin D's skeletal effects. But even here, "outside the setting of nutritional rickets" the picture is fuzzy. For example, children with clinical rickets are at greater risk of fractures, but that association fades into uncertainty for children with vitamin D deficiency but without signs of rickets. The mixed results of studies lead to the inevitable complaints about the lack of firm standards on "how to best define vitamin D deficiency," and calls for more and better research, "well-designed pediatric research studies that encompass multiple geographic and socioeconomic groups," and—the gold standard—large-scale randomized prospective studies.[48]

In 2019 *The New England Journal of Medicine* published the results of just this kind of gold standard study. Seven years earlier, a team led by JoAnn Manson, an epidemiologist at Brigham & Women's Hospital, enrolled about twenty-six thousand subjects in a five-year randomized and placebo-controlled study of the possible role for vitamin D and omega-3 fatty acids in the prevention of cancer and cardiovascular disease. The study found that vitamin D supplementation "did not lead to a significantly lower incidence of invasive cancer of any type or . . . major cardiovascular events . . . than placebo." Vitamin D offered some improvement in cancer survival and showed some promise in preventing type 2 diabetes. One of the limits on the study was that it did not include subjects with very low vitamin D levels. The 2019 report acknowledged that "it remains possible that a trial involving persons with extremely low vitamin D levels would show stronger effects on risk." But that hypothesis could not be tested: fortunately, ethical standards for human subject studies had evolved since the days of the rickets experiments by Hess and Stokes, so "maintaining participants in a vitamin D–deficient state and circumventing real-world clinical care for 5 years would be neither ethical nor feasible."[49]

Medical researchers seem to be in general agreement that vitamin D deficiency, defined as a serum vitamin D level under 20 ng/mL, is powerfully correlated to many lifelong health concerns not limited to skeletal health, but there is little consensus on whether or how to ensure that people achieve that minimal level. The uncertainties pile up when higher levels are considered: First, should everyone be encouraged to maintain higher average levels, or just individuals in high-risk categories? And how are those categories defined? How much should society pay for the testing and the supplements to achieve those higher averages? And finally, what is the upper threshold where risks of "vitamin D toxicity" come into play? Fuzzy definitions, overabundant variables, and uncertainty about benefits and risks make setting guidelines for vitamin D intake (let alone enforcing them) nearly impossible.

And, in a world plagued by uncertainty, with a population increasingly risk averse and intent on informing themselves on health matters affecting them and their families, doubt about whom to trust becomes an integral part of the problem. The doubt and mistrust come from many sources, with the harshest voices often coming from within the medical profession itself, as in a 1980 editorial by the famous quack-buster Victor Herbert in *Archives of Internal Medicine* that condemned the "Vitamin Craze" of the day: "There would be no controversy about the role of vitamin use in health and disease if all health professionals and those whose commentary on health matters is addressed to the public adhered to facts rather than promoting sensational anecdotes alleging efficacy, and if they heeded the axiom that, in matters of health, no substance is safe until proved safe, or effective until proved effective."[50] Herbert hits two of the major sources for vitamin skepticism: specialists with an agenda and headline-hungry journalists and health writers. The rise of the internet has only multiplied the cacophony of competing "authorities" and accusations of corrupt intent.

The most popular approach journalists take in explaining to the public these uncertainties about vitamin D's true health benefits has been to heed the Watergate-era mantra to "follow the money!" Rather than untangling or reconciling conflicting findings, health journalists often start by confidently asserting that corporate interests determine scientific outcomes. So the question becomes not "Why is it so hard to find a one-size-fits-all standard?" but "Who stands to benefit from Vitamin D mania?" At the front of the line of suspects, of course, is what, in this sensational style, should be called "Big Vitamin." And it is a big target indeed. Annual vitamin and supplement sales in the United States top $30 billion. Over half of Americans take at least one vitamin daily, and 10 percent take four or more. Vitamin D sales alone bring in over a billion dollars.[51] Raising the definition of "deficiency" from 20 to 30 ng/dL would be a boon to everyone aboard the vitamin D train, from the sheep farmers producing lanolin to pharmaceutical company stockholders. "Big Testing" comes in for its share of suspicion as well. Understandably so: the global market for clinical laboratory services topped out at over $200 billion in 2019, with vitamin D tests contributing about $500 million, and US labs doing over half of those tests. In recent years, Medicare paid over $300 million annually for vitamin D tests.[52]

The economic critique need not be heavy handed. How a journalist frames a "both-sides" argument can do the work, as in a recent back-and-forth about optimal levels published in *Better Nutrition*. It starts with the standard agreement defining deficiency as a level under 20 ng/mL, then offers a neutral voice on the uncertainty about slightly higher levels: "Many experts consider that 40 to 80 ng/mL is good for overall health." The subtle accusation of economic interest comes in the next two sentences about much higher levels: "The Vitamin D Society . . . goes so far as to say that 100 to 150 ng/mL is an ideal range for whole-body health. However, other medical experts believe vitamin D levels over 150 ng/mL are dangerous."[53]

But in many cases, the muckraking is the thing, as in a 2018 *New York Times* takedown of endocrinologist Michael Holick, an indefatigable advocate for the prohormone.[54] *Medline* lists him as author or coauthor on over three hundred papers on vitamin D, including fundamental biochemical and physiological studies in the early 1970s that established how the skin produces the building blocks of vitamin D; Holick's publications appear in the footnotes of more than one of this book's chapters.[55] But the *Times* dismisses his "book-length odes" and "scholarly articles about a 'vitamin D deficiency pandemic,'" focusing instead on his "extreme" enthusiasm, his quirky musings about whether rickets killed the dinosaurs, how he "elevates his own levels of the stuff with supplements and fortified milk," and his failure to wear sunscreen when biking. But, we are assured, he is no harmless eccentric. No, he "perhaps more than anyone else is responsible for creating a billion-dollar vitamin D sales and testing juggernaut," the result of "mainstream doctors and wellness gurus alike" embracing his message.

But was that message inspired by science or self-interest? The answer seems clear, since he promotes "practices that financially benefit corporations that have given him hundreds of thousands of dollars—including drug makers, the indoor tanning industry and one of the country's largest commercial labs." Equally damning, his interpretation of vitamin D's benefits had been adopted by Gwyneth Paltrow, Dr. Oz, Oprah Winfrey, and other denizens of "the wellness-industrial complex." To be sure, Holick's long history of hearty advocacy and self-promotion made him a target for this criticism.[56] Again and again I found myself reading a scientific study that seemed a tad too cocksure of vitamin D's benefits only to find, sure enough, his name among the authors. His personal website hails him as "The Leading Authority on Vitamin D" and directs viewers to buy his new book that promises "a 3-step strategy to cure our most common health problem" and includes a foreword by Andrew Weil, MD. Although Holick does not market or even endorse any specific brands of supplements—his website does not have a link to a "Dr. Holick's Vitamin D Store"—he clearly is himself a brand, a twenty-first-century Dr. De Jongh.[57]

Oddly, the media seems uninterested in turning the question around: Who stands to benefit from *downplaying* vitamin D's impact, and have they been putting their thumbs on the scale of clinical research? It makes no sense to look for some "Big Skin Cancer Prevention" or a secret cabal that benefits from lower vitamin D levels. But there is a class of actors who stand to benefit from our devoting less attention to monitoring vitamin D status: those who *pay* for all the tests and doctor visits and prescription vitamin D supplements (a rarity as most vitamins are sold over the counter without a prescription). The last paragraphs of the *New York Times* exposé point without pointing to these beneficiaries. A recent study by Excellus BlueCross BlueShield of Rochester, NY, had highlighted "the overuse of vitamin D tests." Excellus had spent "$33 million on 641,000 vitamin D tests" in 2014, "an astronomical amount of money" an Excellus vice president exclaims, considering that their study found that 40 percent of those tested "had no medical reason to be screened." The article concludes with the acknowledgement that the insurer was unable to "rein in the tests," because, in the insurance VP's words, "the medical community is not much different than the rest of the world, and we get into fads."[58]

To be sure, criticism of vitamin fads has been around since the vitamins were discovered, and those critiques fell right into the venerable tradition of fighting quacks and snake oil salesmen. But up to the last few decades, the suspicion was directed at scoundrels marketing unscientific or dangerous products, or drug companies overselling their products' benefits without proper testing. Science was the solution to the problem, not part of the problem itself.[59] Critics could rely on the public's deep faith in science—if vitamin marketers' bogus scientific claims were a problem, the state would regulate them, guided by objective science built upon sound medical studies. Today that faith is all but drowned in suspicion that all science, despite any appearance of objectivity, is bought and paid for by special interests.[60]

: :

The slippery notion of normal vitamin D levels and how best to maintain them is complicated further when we try to factor in the messy consequences of human evolution, migration to extreme latitudes, skin color, diet, and bone calcium metabolism. The Inuit and other people living far from the equator pose a particularly thorny set of paradoxes. Reading through the sources from the early vitamin D era, it is striking how often the subject of "Eskimos" appears, as in this lecture by Harvard pediatrician Joseph Garland in 1925: "The Eskimo lives for half of the year entirely deprived of the rays of the sun, but he subsists on a diet abundant in natural fats and oils, and his children do not have rickets."[61] Archaeological and documentary evidence confirms that, prior to the second half of the twentieth century, rickets was rarely found in Inuit children. Most observers credited the traditional Inuit diet for the persistence of dark skin. "Because of his diet of anti-rachitic fats," anthropologist Frederick Murray explained in 1934, "it has been unnecessary for the Eskimo to evolve a white skin in the sunless frigid zone. He has not needed to have his skin bleached by count-less centuries of evolution to admit more antirachitic sunlight. He probably has the same pigmented skin with which he arrived in the far north ages ago."[62]

Being convinced of the social determinants of health, I was not surprised to learn that rickets became more of a problem in arc-tic communities as they adopted more typical American diets—a transformation begun only a few generations ago. It was intriguing to learn that, while their vitamin D levels had fallen recently, those levels had *always* been lower than average, and that their rates of rickets and other vitamin D–associated health problems are lower than their vitamin D levels would predict. The dietary change was dramatic but was less about giving up walrus blubber or whale oil than traditional interpretations have it. Those foods dominated only the diets of coastal peoples, but rickets was rare in Inuits liv-ing close to the coast and those living far inland. The diet of these circumpolar peoples, little changed in over ten thousand years, was indeed "abundant in natural fats and oils," but not particularly high in vitamin D. Through hundreds of generations, strong selective forces shaped vitamin D and calcium metabolisms that did not rely on vitamin D levels considered "normal" by "Western" stan-dards. Adopting southern foodways took less than a generation; replacing the traditional high-protein, high-fat diet with cereals and sugar upset that ancient metabolic balance. Changes in nurs-ing and infant-feeding practices imposed a direct risk of rickets. Traditionally, Inuits and other Native circumpolar people post-poned weaning well into the third year, advantageous less for vi-tamin D than for the benefits of the mother's "country food" diet, high in protein and fat, low in carbohydrates. Today Inuit infants typically nurse for under a year, taking in milk from a mother who grew up on a "modern" diet; or they are formula-fed and weaned to a cereal and cow's milk diet. Public health programs to reduce rickets focus on vitamin D supplementation, prompting a recapitulation of the arguments about defining "deficiency," "sufficiency," and risks of vitamin D overload in a population primed by evolutionary forces to thrive with "low" levels of vitamin D.[63]

A similar paradox appears at play with vitamin D levels in African Americans: despite much lower average vitamin D levels, their risk of falls and fractures is much lower than white Americans with similar vitamin D status, raising concerns that efforts to increase vitamin D levels in African American adults may have no benefit, and may in fact have negative health outcomes unrelated to bone health.[64] A 2013 study of about two thousand Black and white adults in Baltimore found that, while African American subjects tended to be deficient in vitamin D, markers for how their bodies were utilizing the vitamin were close to normal. The study's leader concluded that "the definition of vitamin D deficiency needs to be rethought" to prevent its overdiagnosis in Black patients.[65] And while further research challenged that study's methods, a growing consensus finds that the paradox is real and significant, complicating the already-robust debate as to how aggressively to push for vitamin D supplements, or promote dietary and lifestyle changes. Accounting for the metabolisms of people whose ancestors had adapted—through thousands of years of natural selection and cultural practice—specific embodied relationships to the sun, to foods, and hence population-specific vitamin D requirements, has never been so important, or so difficult to do.

: :

Which brings us to the clinic. All of these factors—the physiology of vitamin D; the natural selective processes giving rise to lactose tolerance and partial erasure of melanin in northern European "white" people, or to arctic people's more efficient use of calcium and vitamin D; and on top of that all the social factors—taken together make the clinical encounter . . . interesting. The Inuit teenager presents with "insufficient" vitamin D levels. But what are her actual risks, and what remedies (if any are required) are most advantageous? A woman whose family emigrated from Pakistan two years ago enters the clinic clad in traditional full covering; she is in her third month of pregnancy. A man's intake form describes him as "Black or of African Descent." How does the clinician parse that to address a finding of vitamin D level twice the level considered "normal"? A sixty-year-old "white" construction worker has been megadosing vitamin D capsules because he read that it could prevent colon cancer, which killed his father. His bloodwork shows signs of hypercalcemia. Each of these patients presents unique diagnostic issues not answered by any single metric. So much training and clinical protocol is based on making, then acting on, binary racial distinctions. More effective clinical practice starts from recognizing that patients don't line up neatly by color or race and calls for taking deeper personal, social, and family histories, even though that care may serve only to guide the clinician in deciding which sanctioned binary racial or color-based protocols to follow.

Are clinicians today then stuck between the Scylla of racial essentialism and the Charybdis of "color-blind" vitamin D status assessments? Atlanta pediatrician Norman Carvalho concluded his impactful 2001 report on the uptick of rickets cases he was seeing with a promising third way. He begins with the conventional wisdom: "Individuals with darkly pigmented skin require more sun exposure to synthesize vitamin D, thus, placing this dark-skinned patient at higher risk." But he continues: "Secular lifestyle changes may also decrease the amount of time infants and children are outdoors during sunlight hours. These include factors such as both parents working long hours, increased sedentary indoor lifestyle . . . poor outdoor air quality . . . with available indoor air conditioning."[66] We want to put these modern cultural and social factors into context with the historical factors that produce rickets in African Americans disproportionately: factors such as poverty, crowded urban environments with limited solar exposure, attitudes about breastfeeding initiation and duration, the likelihood of weaning to low-calcium diets that include fewer vitamin D dairy products, distrust of the medical establishment's pronouncements about diet and supplementation—distrust of the medical establishment, period! In other words, for African Americans, the factors leading to vitamin D deficiency are many, and most are inextricably linked to the legacy of slavery. These may be social determinants, but many will be read in the clinical setting along racial lines.

There is a growing call among health professionals and social scientists "to move past the use of race as a tool for classification in both laboratory and clinical research." This is no naïve demand that biomedicine ignore how frequently given traits appear within or between racially or ethnically defined groups or use these data to guide clinical practice; instead, these critics seek an escape from the paradox of relying on "race" or ancestry as "a tool to elucidate human genetic diversity," knowing all the while that race "is a poorly defined marker of that diversity."[67] In theory science can identify biologically significant differences across socially defined "racial" categories, but in reality those categories leak at every seam, and so should not be treated as coherent guides for clinical decisions about individuals.

The example of vitamin D does not offer an escape from the paradox. But it stands as a golden example of why resolving it is crucial. Consider how we set standards for low, adequate, and high levels of vitamin D. Part of the difficulty is how a normal range is determined, described by Michael Holick as "obtaining blood from several hundred volunteers and deeming them to be normal and to perform the measurement of the analyte and do a distribution with a mean ± 2SD as the normal range."[68] Of course, Holick is critical of this approach because he is certain that today's mean, and hence today's "normal," is unhealthily low. That may well be, but it should be noted that in this simple example, "normal" also ends up skewing "white." Researchers testing new potent vitamin D supplements in the 1920s and 1930s were sensitive to the risk of standards skewing "Black" and sought to repeat all-Black studies with children "of the dominant race" or reported whether an intervention that prevented rickets in white children also sufficed in Black subjects.[69]

In short, setting standards without accounting for differences across identified population groups risks perpetuating racial disparities in health outcomes, because those standards don't treat to the mean of the population most affected by rickets, but to the "racial norm"—which is to say "white." Some of these miscues are essentially physiological: failure to distinguish differences in how vitamin D and calcium are utilized, the UV screening effect of melanin, or differences in skin cancer risk. Other factors are culturally and historically rooted within identified "racial" groups: pertinent examples include breastfeeding practices, infant swaddling, sun avoidance, full body covering, and diet. Diet of course, can combine the physiological and the cultural—the Inuit's "country foods," the traditional fatty fish-rich diets of coastal peoples of the far north, or dairy consumption.

From the start of the vitamin D era, American health experts promoting universal fortification saw milk as a nearly "automatic method," since, as Hess put it in 1933, milk provided both calcium and phosphorus, and "since every child is fed on milk."[70] Dairy products can only be the perfect vitamin D delivery vehicle for communities who continue to consume milk after weaning, a cultural choice undoubtedly linked to the biological trait of lactose tolerance. Only one third of humans are lactose *tolerant*—that is, they maintain into adulthood the ability to digest lactose.[71] For the majority, consuming dairy can result in bloating and uncomfortable symptoms of the body fighting to digest the undigestible. Maldigestion studies since the 1980s estimate that almost half of Americans are lactose intolerant—with upward of three-quarters of African Americans and Native Americans, and approximately half of Mexican Americans, unable to consume dairy without risk of lactose maldigestion.[72]

Not everyone is convinced of these high rates of lactose intolerance, arguing that the studies establishing those rates fed test subjects far more lactose than found in a typical meal. These tests are good at identifying individuals who do not produce lactase but cannot predict if modest dairy consumption will make them sick.[73] Many health professionals question whether someone's inability to fully digest dairy should prevent them from enjoying the "perfect food," albeit in smaller servings. Some go so far as to question if lactose intolerance is a myth, casting "dairy avoidance" as a greater threat to health.[74] Inevitably the "myth" they have in mind is not the fact of lactase nonpersistence in most humans, but the common belief that those who don't easily digest lactose must abstain completely from lactose-laden foods. An internet search on "lactose intolerance" produces a glut of helpful posts that lay out basic facts and statistics. A sizable portion of these posts then turn to a two-part "common sense" conclusion: dairy is a crucial source of calcium, phosphorus, and vitamin D for bone health and metabolism; and those unpleasant symptoms of lactose intolerance can be avoided if dairy is consumed with care and in moderation. Readers are also directed toward lactose-free dairy products and lactase enzymes to promote digestion of dairy.

:: ::

The ghost of the long-shuttered dairy refinery we visited in the introduction should remind us that our relationship to milk is changing, and that perhaps its strained and artificial relationship to vitamin D should as well. Regardless of how many Americans have trouble digesting lactose, it is time to revisit the notion of using milk—dairy, soy, coconut, or oat—as an "automatic method" for delivering vitamin D. The marriage of Elsie and the big "D" does not need to be held together "for the kids." Better yet, we should reconsider the whole union of food industry and pharmaceuticals, at least for vitamin D, compromised by false promises of one accurate and healthy dosage for all. There are better alternatives for the future.

9

Epilogue

Hard Choices, Slow Violence

Perhaps the way to escape uncertainty about vitamin D requirements is clinical genomics, with its promise of unheard-of diagnostic specificity: a patient's gene sequence will provide clinicians all the nuance and precision they could ask for. Instead of relying on a patient's self-reporting of their ancestry, the physician will skip the entire tangle of presumed "race traits" and pull up the patient's whole genome sequence to identify individual markers for skin cancer risk, lactose tolerance, and factors affecting their calcium metabolism.[1] Genome-guided "precision medicine" promises ultramodern tools to revive the premodern ideal of focusing on the "specificity" of each patient: deep interrogation and observation of the whole patient in their environment, tailoring diagnosis and treatment to the specific patient's needs—in short, treating specific patients, not specific diseases.

But we cannot sit still. While we wait for a *Deus ex Genoma* to settle things, there are issues we can address. In the clinical encounter, taking deeper personal, social, and family histories can steer practitioners away from overreliance on binary race- or skin-pigment-based protocols. We need to reduce uncertainty about vitamin D—how little is too little, how much is too much, and how do we get everyone to their personal optimal level? As with so much nutritional epidemiology, every year seems to bring new rafts of contradictory studies about the risks of vitamin D deficiency.[2] Simply defining "deficiency" is still a challenge. Ironically, one way to confront the dizzying variables to be accounted for in addressing vitamin D deficiencies in a diverse population in the United States is to embrace a truly global perspective on standard setting, with more cross-national comparisons.[3]

But why wait? There is an option, as obvious as the sun in the sky, that all but assures adequate vitamin D without spending billions of dollars on vitamin pills, or taxing the medical system with excess costly assays, reducing in turn the risk of pharmaceutical and commercial diagnostic companies corrupting the science, and which employs a built-in regulatory system to avoid vitamin D overload. A century ago, Los Angeles physician Robert Ramsay observed that rickets in his city was "of a milder type" than in eastern towns, and that it affected "all classes of society." Unlike back East, race and class were poor predictors in Los Angeles because "sunlight and fresh air are so generally available for rich and poor alike."[4]

Sunlight, then as now, is free, plentiful, and, as Jonas Salk famously observed in a different context, not subject to patents or licensing fees. If not for its one great risk, it could eliminate most of the uncertainties about vitamin D deficiency. Adequate sun exposure is the "natural" cure for low vitamin D, and in the early years of the vitamin D era, heliotherapy played an important role, together with cod liver oil, in promoting healthy bones and fighting infections. But since the 1960s as rickets continued its long slide out of public discourse, skin cancer—presumably a far greater direct health threat than vitamin D deficiency—jumped to the forefront.

Guidelines for preventing skin cancer recommend avoiding direct sun exposure. Does this mean that to be safe we must avoid the sun and get our "D" from diet and supplements? "Yes!" according to the Skin Cancer Foundation, which argues that "while a limited amount of vitamin D can be obtained from exposure to the sun's ultraviolet (UV) radiation, the health risks of UV exposure—including skin cancer—are great," and suggests that we should get the vitamin "from sources like oily fish, fortified dairy products

and cereals, and supplements." "No!" say epidemiologists convinced of the dangers of vitamin D deficiency. A Dutch preventive
medicine specialist, after conducting an umbrella review of studies on the topic, argued against supplements and in favor of the
sun, suggesting that, without clear standards for supplementation, people should maintain healthy vitamin D levels "with an adequate diet and 30 minutes of sunlight twice a week."[5] This strong
cultural ambivalence is understandable. Skin cancer falls squarely
in the category of modern environmental or lifestyle diseases, but
effective proposals to prevent it run headlong into a deep vein of
faith in the health-giving properties of the sun. And, of course,
the uncertainties concerning skin cancer and the relationship between sun, vitamin D, and health bring the focus back to skin, notably skin pigmentation, which brings us back to questions of race.

We need to develop and publicize standards for safe sun exposure. Sensible, easy, and safe heliotherapy is still a great way of letting the body provide and monitor its vitamin D needs. But apart
from that, fear of the sun is a major driver in pushing Americans
to shun outdoor activities and nature contact, proven sources of
bountiful health and social benefits.[6] Finally, so that we can know
its true prevalence, every case of rickets should be reported to
public health agencies. Much of the "surprise" that attended the
resurgence of rickets in the early twenty-first century—if a resurgence is what it was—could have been avoided, as could many of
the cases, had clinicians been aware earlier of new circumstances
driving an uptick in the incidence of rickets.

After reading this book, one may be excused for wondering if
rickets remains a nonreportable condition not simply because its
incidence is believed to be so low but because it is still believed to
afflict mostly the poor and nonwhite—that rickets was and is a
troubling but inevitable consequence of intractable poverty and
structural racism. In this formulation, children with rickets will
likely remain, in writer and activist Rob Nixon's formulation, "the
casualties most likely not to be seen, not to be counted," invisible
victims of the process he calls "slow violence."[7] Nixon applied the
framework of slow violence to such familiar environmental justice

issues as persistent toxic environmental hazards, climate change, and structural legacies of war, but the concept has proven flexible enough to be applied to the coronavirus pandemic and persistent health care inequalities, and I would argue that it applies equally well to vitamin D deficiency. Consider a partial list of traits associated with slow violence: it is persistent, sometimes spanning generations; it is chronic rather than being defined by a clear episode; it is usually imperceptible as it happens and vague in both its sources or the forces behind those sources; it has a disproportionate impact on women and children; and it is "not just attritional but also exponential, operating as a major threat multiplier."[8]

Some forms of slow violence, especially those related to toxic environments or products, are abetted by the polluters' ability to "manufacture doubt" about the nature and scope of these threats—a process historian Robert Proctor named "agnotology." The doubts about vitamin D deficiency—its causes, its uneven and unpredictable impact across populations and places—do not lend themselves to theories of corporate malfeasance in the same way as do other toxic threats. Historians rummaging through the archives of vitamin D's history will search in vain for smoking-gun proof of back-room conspiracies to hide "the truth about rickets." That does not mean there was no malfeasance or obfuscation, but that the blame is society-wide and deeply historical, the costs of accepting responsibility and making corrections have all along been deemed unacceptable. Writing of climate change and other slow assaults on the environment, Geologist Jill Schneiderman captured the resulting impotence and stasis: "Human time flows in brisk countercurrent to slow-motion environmental time, and guarded agents in lethargic systems listlessly search for avenues of action, knowing not in what direction to move." As a result, she suggests, the causes of slow violence can "masquerade as inevitable."[9] The long history of vitamin D deficiency and rickets this book relates might be mistakenly read to support the notion that rickets was not merely masquerading but was in fact inevitable.

"Inevitable" suggests a lack of agency, of choice. But new circumstances giving rise to rickets, it must finally be asserted, are always choices. I wanted this book to examine the notion of whether rickets might be a clear example of a "race-specific" disease, and at this point readers should have a clear idea about the evidence that has made that seem a plausible notion. But a fuller reckoning leads inexorably to the opposite conclusion, that rickets is one of the clearest examples we have of a truly socially constructed disease. If it looks like a "racial" disease, that is because our choices have made it seem so. A century ago, when Lawson Dick wanted to show that rickets was a product of modernity—industrialized civilization to be specific—he could not avoid discussing the Greek-born physician Soranus, who made clear that rickets was familiar in second-century CE Rome. The rickety children of Soranus's Rome, like those of Dick's London, were not victims of civilization, but of unhealthy choices about how civilization's citizens should live.

The infamous eruption of rickets in eighteenth- and nineteenth-century Britain was not an inevitable consequence of "the Industrial Revolution" as much as the weight of thousands of decisions shaping the lived experience of industrialization. Spinning jennies and power looms do not drain a body's stores of vitamin D, nor does the transition of an agrarian people to wage-earning city dwellers. But the captains of capital and the landlords of Dickensian Britain *chose* to pay impoverishing wages and house the families of millworkers in sunless, crowded slums, their children starved of food and starved for light. Likewise, forcibly transporting millions of Africans to latitudes far north of their ancestral homes did not produce the disproportionate burden of rickets their children and children's children suffered. It was not a racial "latitude mismatch" that produced the common sight of bowlegged Black babies in southern slave labor camps; it was thousands of decisions by their captors to starve enslaved infants and then turn a blind eye to the results.

The same pattern holds for the migrants to northern cities of the United States in the nineteenth and early twentieth centuries, and the northward migrations from the former colonies of European empires. Crushing rates of rickets among African American and Italian immigrants a century ago were not the product of a "latitude–skin color mismatch," but the result of confinement in some variation of New York's five-floor railroad tenement with its twenty-five by one hundred footprint and dark and miasmatic interior; the epidemic of rickets was the result of a thousand other choices driven not by race, but by racism.[10]

America's "cure" for rickets was a choice as well. Steenbock and the other "vitamin hunters" a century ago gave us new options for preventing and curing rickets. Armed with the knowledge of vitamin D's intricate relationship with ultraviolet light, we could end rickets by engineering the built environment and our place in it, or we could accept life in the coal smoke and shadows and embrace the modern biomedical "fix" on offer from the licensees of the Wisconsin Alumni Research Fund. It was an offer we *could* refuse, as other nations did. But we chose the path where a multibillion-dollar industry provides the vitamin-infused sheep sweat we add to the milk that another multibillion-dollar industry tells us is the "natural" source for minerals and vitamins—a source that most humans have done fine without for thousands of years.

There were worse options. We could have chosen to live without the sun at all, dose ourselves with synthetic vitamin D, and move into climate-controlled homes, cars, stores, offices, schools, and factories, kept safe from a dangerous natural environment by high-efficiency particulate air filters and UV-filtering windows, the only ultraviolet light in our homes that beam, unseen within the antimicrobial sanitizer module at the heart of every HVAC system. Of course, we have moved quite a few stops toward that destination—we were well underway already in 1968 when microbiologist René Dubos warned that "the largest part of life is now spent in an environment conditioned and often entirely created by technology." Dubos found it "significant and disturbing" that our contacts with "the rest of creation are almost always distorted by artificial means." Modern humans were alienated, not only from each other, "not only from nature, but more importantly from the deepest layers" of their fundamental selves.[11] Many years later, and further down the path to our encapsulation, choices remain. Having a biotechnological "cure" for vitamin D deficiency does not mean we must rely on it exclusively. We can have sound bones and bright homes, healthful foods, and the benefits of close contact with the natural world. We need not turn our backs to the sun.

Acknowledgments

Years ago, the only plans I had for writing about rickets cast it in an important but humble role in what was to be my second book. Rickets was to be an object lesson in a longish history of Americans and their built environment. But the more I wrote about rickets and the more I talked about it with friends and colleagues, the clearer it became that the story of rickets and vitamin D needed to be far more than a pithy example in another book, that it deserved to be its own story, with pithy examples of its own. And so I readjusted my long-term writing schedule and committed to writing a book about rickets . . . as soon as I completed the book I was working on. But that is not what happened, and for bumping this project from "third book" to "next book" (and for so many other kindnesses), I am happy to thank David Troyansky. David was the chair of the Brooklyn College History Department when I joined the faculty. During one of my annual performance reviews, he observed how enthusiastic I was getting about rickets. He suggested running with the momentum I was building and pushing "the rickets book" to the front burner. At that point I decided *Starved for Light* would be my second book.

Initially the plan was for this to be one of those "just-the-facts" monographs, somewhere between the "very short introduction" and slightly longer "biography of disease" genres. If you're holding the physical book, you know that didn't happen. Some of the inspiration for this added heft came from colleagues who came to talks in which I expounded the many connections between rickets and history. Their kind attention and great ideas added corrections, context, and inspiration for a study that needed a skosh more room than I'd originally planned on. In addition to the many helpful audience members who had questions, comments, and comments-more-than questions, I gratefully acknowledge co-panelists or organizers: M. Allison Arwady, Roberta Bivins, Janet Golden, Raechel Lutz, Sarah Pripas-Kapit, Sally Romano, Kendra Smith-Howard, and Helen Zoe Veit. And I'm grateful to the organizations that hosted these sites of intellectual incubation: the American Society for Environmental History, International Consortium of Environmental History Organizations, New York Metropolitan Seminar for Environmental History, and the Society for the Social History of Medicine. Special thanks go to my colleagues and friends in the American Association for the History of Medicine. AAHM has provided so much more than conferences and publications: from its members I have received bountiful guidance, inspiration, and friendship.

Many friends and colleagues made unique contributions along the way. Some I met way back in graduate school: Jack Davis, who this time round didn't give me my book title, but read chapters and returned them filled with pithy and constructive marginalia; Nicholas Bloom, who invited me to present some of my research at a conference he organized on "Urban Childhoods"; and Martha Gardner, who, during many conference meetups over coffee or dinner, listened patiently to my ideas and shared her insights. My first graduate school professor, Ray Arsenault, provided sage advice and moral support at a time during this project when I really needed it.

My medical- and environmental-history friends have been major sources of ideas and guidance: special thanks to Roberta Bivins, Allan Brandt, Ted Brown, Lara Freidenfelds, Carla Keirns, Susan Lederer, Beth Linker, Neil Maher, Leslie Reagan, Susan Reverby, Jeffrey Reznick, Naomi Rogers, Walt Schalick, John Swann, Arleen

Tuchman, Jackie Wolf, and Michael Yudell. Janet Golden has been an especially good friend to me and this project, providing gleanings from her research, opportunities to publish (and cowrite) some think pieces, and helping organize a conference panel. I also had advice from professionals who are not professional historians, including anthropologist Meredith A. B. Ellis, who shared insights from her study of the remains buried under the Spring Street Presbyterian Church; physician and epidemiologist Bruce Lanphear, who has over the years provided advice about pediatric health; and my good friend Carol Ford, who gave advice, listened to my ideas, and, in her role as editor of a medical journal, offered a place to get some of my ideas about rickets out to a nonhistorian audience. My Brooklyn College colleagues have been supportive of me as I pursued what must seem an idiosyncratic study, but special thanks go to my long-suffering office mate Michael Rawson, who has endured many a breathless tale of yet another "amazing discovery" about rickets, many a complaint about elusive sources, and many a joke or song about rickets or cod liver oil, some that would never appear in an academic book.

Many librarians provided invaluable assistance in researching this book. First and foremost are the brilliant and skilled professionals I worked with at the New York Academy of Medicine. Arlene Shaner helped find sources ancient and modern; Constance Malpas spent many hours discussing both this project and the next. Other colleagues there, notably Patricia Gallagher and Winifred Capowski-King, were both supporters and friends. I also received guidance, and carts full of research materials, from the librarians and archivists at many collections, including the Schlesinger Library in Harvard's Radcliffe Institute; the National Archives; the Historical Medical Library of the College of Physicians of Philadelphia; the American Philosophical Society Library; and Harvard's Countway Medical Library, with special thanks to Dominic Hall in the Countway's Center for the History of Medicine. At the National Library of Medicine's History of Medicine Division, Steve Greenberg and Paul Theerman offered advice and an opportunity to give a seminar lecture early on in this project. In recent years I have come to rely heavily on the resources provided by the librarians at Brooklyn College of the City University of New York.

I am happy and proud to call Brooklyn College my institutional home, a college with brilliant and supportive colleagues, bright, knowledge-hungry, and creative students—a student body among the most diverse of any campus in the US. It's a school determined, despite being besieged by funding and infrastructural challenges so common in public higher education today, to continue fulfilling its historical mandate as a stepping stone for the city's working class and immigrant families. This project benefitted directly from a generous book completion award from the CUNY Office of Research.

Brooklyn College and CUNY also provided access to a universe of research resources that were crucial in producing this book. It does not diminish the work of librarians and archivists I met in person to note that so much historical material is now available without travelling great distances to gain access to printed and archival documents. It's a testament to how far this shift to the internet has gone that my file cabinets are stuffed with photocopies I made years ago from physical journals (or microforms!) found in great research libraries, but that my hard drives contain images of many of those same sources scanned (much more clearly) by large-scale projects such as Google Books, the Hathi Trust, the Medical Heritage Library, and JStor. And so, in addition to thanking the librarians I met, I offer heartfelt thanks to the armies of patient technicians who created the microfilms or scanned documents, to the public and private organizations that make historical newspapers available, to those who have advocated for getting full-text "Runaway" advertisements online and full-text searchable, and, of course, to the many librarians who aided in all these projects that allow me to conduct some of my best research in my pajamas in my home office at 2:00 a.m.

My editor at the University of Chicago Press, Karen Merikangas Darling, has been a powerful and patient ally, starting with our first coffee conversation at an AAHM annual meeting, where we discussed not rickets specifically, but the project that then was slated to be my "second book." She did not hesitate in her support when I proposed postponing that project and concentrating on rickets. Perhaps she saw my enthusiasm; perhaps some talks she attended convinced her that there was a book's worth of material in vitamin D and rickets. Over the years I have, no doubt, tested her patience, but you would never guess that from her steady, good-humored manner and gentle nudging when a promised chapter appeared to have gone AWOL. Her comments reined in stylistic excesses, snipped out bad Dad jokes, and reshuffled whole passages, all to enhance the book's readability and analytic bite. As the manuscript moved into production, Fabiola Enríquez Flores provided invaluable guidance and technological support at my end and worked institutional magic at the press's end to move things along. I am very grateful to the two anonymous reviewers whose detailed notes and calls for revisions kept me up more than a few nights, but to good effect.

This is a book about light and about nourishment, how both are crucial to health, even if they seem to share only a tenuous relationship. The mystery of vitamin D was solved by demonstrating how sun and sustenance were inextricably linked, how shortfalls in either weakened the constitution and softened the bones. This relationship always made perfect sense to me because daily life with my family provided a powerful analog to a substance mysteriously provided by both light and nutrition. My life partner, Janis, has provided soul nourishment and corporeal sustenance—the hearty fare of creature comfort and the brightest of light, even in the dark times during which this book was completed. In more mundane matters, Janis and our children have been patient with my obsession with vitamin D and rickets. Janis and Justin read drafts of the manuscript and worked hard to get me to dislodge troublesome prose and clarify muddy thinking; Jeremy provided sound logic in his role as tiebreaker in some of these discussions. Johanna, by her lived example, reminded me to rein in any inclination to celebrate uncritically the progress of the modern medico-industrial complex. Love and family, food and light.

But this book is also about the damage that the want of sustenance or a surfeit of darkness can bring, and the efforts to make straight what deficiency has made crooked. And again, the family, and perhaps the human family, provides a potent analog to that iconic symbol for healing bent bodies—the Andry Tree. But who is the sapling and who the stake? Over a lifetime, over the generations, we have all been one, then the other, proving, or at least striving in hope to prove, that what is crooked can indeed be made straight. I end these acknowledgments by restating my thanks to all who have helped in the effort to make this history straight and true. The distortions and flaws remaining are due entirely to my own deficiencies.

Notes

Introduction

1. James Dudlicek, "Morale Boosters," *Dairy Foods*, May 1, 2006, http://www.dairyfoods.com/articles/print/87099-morale-boosters.

2. The RDA in play in the United States was established by the Food and Nutrition Board of the National Academies of Sciences, Engineering, and Medicine. NIH, "Vitamin D: Fact Sheet for Health Professionals," accessed October 31, 2023, https://ods.od.nih.gov/factsheets/VitaminD-HealthProfessional/. These standards are far from settled, a topic taken up in chapter 8.

3. Alex Matthews-King, "Huge Increase in 'Victorian Diseases' Including Rickets, Scurvy and Scarlet Fever, NHS Data Reveals," *Independent*, February 25, 2019, https://www.independent.co.uk/news/health/victorian-disease-gout-rickets-vitamin-d-mumps-scurvy-measles-malnutrition-nhs-hospital-admitted-a8795686.html.

4. *Coco*, directed by Lee Unkrich, Adrian Molina (Disney-Pixar, 2017).

5. John Howland, "Starving for Sunshine," *The Delineator*, March 1926, 16, 72, 74, quotes 16. On the death statistics, Sister Mary Theodora Weick, "A History of Rickets in the United States," *American Journal of Clinical Nutrition* 20, no. 11 (November 1967): 1234–41.

6. Or, more specifically, a group of steroid hormones. Vitamins proper were defined by Casimir Funk, who found that something, he presumed an amine, in the polishings of rice, cured polyneuritis in pigeons. This *amine*, then, was *vital* to life, a "vital amine." R. D. Semba, "The Discovery of the Vitamins," *International Journal for Vitamin and Nutrition Research* 82, no. 5

(2012): 310–15. Traditional vitamins are organic components that, while not a source of energy, are required in very small amounts in order to promote growth and health. Because the body can manufacture its own vitamin D, it does not meet this definition. But since it *can* be supplied through the diet, it is customary for purposes of dietary policy to classify it with the true vitamins. "Vitamin D" comes in several forms: ergocalciferol (vitamin D2), which is synthesized by irradiating vegetative matter, and cholecalciferol (vitamin D3), the form synthesized in the skin when exposed to ultraviolet B (UVB) rays from sunlight. Either form can be used to fortify foods.

7. This may be just as well; after all, "sunshine steroid hormone" would be a lot harder to sell. For a brief history of the discovery of vitamins, see Kenneth J. Carpenter, "The Nobel Prize and the Discovery of Vitamins," June 22, 2004, https://www.nobelprize.org/prizes/themes/the-nobel -prize-and-the-discovery-of-vitamins/. For a general history of vitamins as commodities, see Rima D. Apple, *Vitamania: Vitamins in American Culture* (New Brunswick, NJ: Rutgers University Press, 1996).

8. Anthony W. Norman, "From Vitamin D to Hormone D: Fundamentals of the Vitamin D Endocrine System Essential for Good Health," *American Journal of Clinical Nutrition* 88, no. 2 (August 2008): 491S–99S; for fair-skinned people, a five-to-fifteen-minute sunbath can produce enough vitamin D for a month; melanin in darker skin filters much of the sun's UV radiation, necessitating more time in the sun for the same benefit. Obviously, latitude, weather, and season affect how much UV will reach the skin.

9. Lea Tremezaygues, "Cutaneous Photosynthesis of Vitamin D: An Evolutionary Highly Conserved Endocrine System That Protects against Environmental Hazards Including UV-Radiation and Microbial Infections," *Anticancer Research* 26 (2006): 2743–48.

10. Frances Sizer and Ellie Whitney, *Nutrition: Concepts and Controversies*, 13th ed. (New York: Cenage Learning, 2013), 244.

11. For example, renal, or hypophosphatemic vitamin D-resistant rickets, usually a genetic disorder, involves compromised ability to process phosphorus and calcium; see Manisha Sahay and Rakesh Sahay, "Renal Rickets: Practical Approach," *Indian Journal of Endocrinology and Metabolism* 17, no. 1 (October 2013): S35–S44.

12. Lewis S. Mace, "Rickets and Proprietary Infant Foods—Report of a Case," *California State Journal of Medicine* 2, no. 7 (1904): 228; N. F. Carvalho, R. D. Kenney, P. H. Carrington, and D. E. Hall, "Severe Nutritional Deficiencies in Toddlers Resulting from Health Food Milk Alternatives," *Pediatrics* 107, no. 4 (2001): E46.

13. The apparent uptick in rickets and vitamin D deficiency prompted several literature reviews, including A. J. Rovner and K. O. O'Brien, "Hypovitaminosis D among Healthy Children in the United States: A Review of the Current Evidence," *Archives of Pediatric and Adolescent Medicine* 162, no. 6 (June 1, 2008): 513–19; and Pamela Weisberg, Kelley S. Scanlon, Ruowei Li, and Mary E. Cogswell, "Vitamin D and Health in the 21st Century: Bone and Beyond: Nutritional Rickets among Children in the United States: Review of Cases Reported between 1986 and 2003," *American Journal of Clinical Nutrition* 80 (2004): 1697S–705S.

14. Rovner and O'Brien, "Hypovitaminosis D," 517.

15. Although in an interview for *USA Today*, Carvalho made less of the distinction, saying that his patient's rickets was due "simply because their parents fed them soy- or rice-based beverages that did not contain vitamin D, instead of cow's milk." " 'Health Food Drinks' Spur Rickets Revival," *USA Today*, October 2001, 3.

16. Suzanne White Junod, "Folic Acid Fortification: Fact and Folly," accessed October 31, 2023, https://www.fda.gov/files/about%20fda/published/Folic-Acid-Fortification--Fact-and-Folly.pdf. This article originally appeared in the "Making History" column of the issues 4, 5, and 6 of the 2001 series of *Update*, the bimonthly publication of the Food and Drug Law Institute; reprinted in Meredith A. Hickman, ed. *The Food and Drug Administration* (Hauppauge, NY: Nova Science, 2003), 55–62.

17. Daniel Freund, *American Sunshine: Diseases of Darkness and the Quest for Natural Light* (Chicago: University of Chicago Press, 2012).

18. Theobald A. Palm, "The Geographical Distribution and Aetiology of Rickets," *The Practitioner* 45 (1890): 270+, 341; Edward Carpenter, *Civilization: Its Cause and Cure, and Other Essays*, 14th ed. (London: George Allen & Unwin, 1916), 26–27.

19. For example, see Allan M. Brandt and Martha Gardner, "The Golden Age of Medicine?" in Roger Cooter and John Pickstone, eds., *Medicine in the Twentieth Century* (Netherlands: Harwood Academic, 2000), 21–37.

20. For example, see chapter 4, "Deficiency Diseases," in Leonard F. Peltier, *Orthopedics: A History and Iconography* (San Francisco: Norman, 1993), 85–99; or Steven B. Beck, "Rickets: Where the Sun Doesn't Shine," in Kenneth Kiple, ed., *Plague, Pox and Pestilence* (London: Phoenix Illustrated, 1999) 126–35. For Steenbock, see Kevin A. Walters, "Before the Foundation: Harry Steenbock and the Patenting of University Science, 1886–1925," (PhD diss., University of Wisconsin, 2019).

21. Apple, *Vitamania*; also Rima D. Apple, "The More Things Change: A Historical Perspective on the Debate over Vitamin Advertising in the United States," in John Ward and Christian Warren, eds., *Silent Victories:*

The History and Practice of Public Health in Twentieth Century America (New York: Oxford University Press, 2007), 193–206; and Susan E. Lederer. *Subjected to Science: Human Experimentation in America before the Second World War*, Henry E. Sigerist Series in the History of Medicine (Baltimore: Johns Hopkins University Press, 1995).

22. Roberta Bivins, "Ideology and Disease Identity: The Politics of Rickets, 1929–1982," *Medical Humanities* 40, no. 1 (June 2014): 3–10, https://doi.org/10.1136/medhum-2013-010400; Roberta Bivins, "'The English Disease' or 'Asian Rickets'? Medical Responses to Postcolonial Immigration," *Bulletin of the History of Medicine* 81, no. 3 (Fall 2007): 533–68; M. Allison Arwady, "The Uses of Rickets: Race, Technology, and the Politics of Preventive Medicine in the Early 20th Century" (MD thesis, Yale University School of Medicine, 2008); Freund, *American Sunshine*.

23. Janet Golden, *A Social History of Wet Nursing in America: From Breast to Bottle* (Cambridge: Cambridge University Press, 1996); Jacqueline Wolf, *Don't Kill Your Baby: Public Health and the Decline of Breastfeeding in the 19th and 20th Centuries* (Columbus: Ohio State University Press, 2001); Richard Meckel, *Save the Babies: American Public Health Reform and the Prevention of Infant Mortality, 1850–1929* (Ann Arbor: University of Michigan Press, 1990); Helen Zoe Veit, *Modern Food, Moral Food: Self-Control, Science, and the Rise of Modern American Eating in the Early Twentieth Century* (Chapel Hill: University of North Carolina Press, 2013); on milk, see Kendra Smith-Howard, *Pure and Modern Milk: An Environmental History Since 1900* (Oxford: Oxford University Press, 2013).

24. Kenneth Kiple and Virginia Himmelsteib King, *Another Dimension to the Black Diaspora: Diet, Disease and Racism* (Cambridge: Cambridge University Press, 1981); Richard H. Steckel, "A Dreadful Childhood: The Excess Mortality of American Slaves," *Social Science History* 10, no. 4 (Winter 1986): 427–65; Susan E. Klepp, "Seasoning and Society: Racial Differences in Mortality in Eighteenth-Century Philadelphia," *William and Mary Quarterly* 51, no. 3 (1994): 473–506.

25. Marie Jenkins Schwartz, *Birthing a Slave: Motherhood and Medicine in the Antebellum South* (Cambridge: Harvard University Press, 2006).

26. Jerome C. Rose, ed., *Gone to a Better Land: A Biohistory of a Rural Black Cemetery in the Post-Reconstruction South*, Research Series, no. 25 (Fayetteville: Arkansas Archeological Survey, 1985); Leslie M. Rankin-Hill, *A Biohistory of 19th Century Afro-Americans: The Burial Remains of a Philadelphia Cemetery* (Westport, CT: Bergin and Garvey, 1997); A. C. Aufderheide and C. Rodriguez-Martin, eds., *The Cambridge Encyclopedia of Human Paleopathology* (Cambridge: Cambridge University Press, 1998).

27. "Farmland Dairies to Close Wallington Plant, Lay Off 323," nj.com, December 19, 2013, https://www.nj.com/business/2013/12/farmland _dairies_to_close_wall.html. Only a few months earlier, Consolidated Dairies, just a mile from the Farmland plant, had shuttered its doors. The small distributor began in 1923 and in its last year delivered to over two thousand locations in the New York. Linda Moss, "Wallington's Consolidated Dairies Closes Shop," September 30, 2013, NorthJersey.com; http://www.northjersey.com/news/ 225889621_Wallington_dairy_closes _shop.html.

28. Farmland was owned by Borden Milk Products, in turn owned by Grupo Lala, a Mexican dairy company. Consumption was 31.38 gallons per person per year in 1968 and fell to 15.6 gallons per person per year in 2021. For 1968, see Judith Putnam and Jane Allshouse, "Table 13: Fluid Milk and Cream: Per Capita Consumption, 1968–89," *Food Consumption, Prices, and Expenditures, 1968–89* (Washington, DC: United States Department of Agriculture, Economic Research Service, 1991), 41; for 2021, USDA Economic Research Service, "Dairy Data," accessed November 1, 2023, https://www.ers.usda.gov/data-products/dairy-data.

Chapter 1

1. Architectural historian Bainbridge Bunting said of the building: "Countway represents the best and most thoughtful application of the bombastic idiom of the late 1960s and early 1970s." Bainbridge Bunting, *Harvard: An Architectural History* (Cambridge, MA: Harvard University Press, 1998): 257.

2. The Warren Collection originated as the lifelong project of the eminent surgeon John Collins Warren (1778–1856), who donated his collection to Harvard University when he resigned from the Medical School in 1847. Warren Anatomical Museum Collection, Countway Library, Harvard University, accessed November 2, 2023, https://countway.harvard.edu /center-history-medicine/collections-research-access/warren-anatomical -museum-collection.

3. J. B. S. Jackson, *A Descriptive Catalogue of the Warren Anatomical Museum* (Boston: Williams, 1870), 307.

4. George Cheyne Shattuck Jr. paid 300 francs for the skeleton and another 150 francs for six plaster casts of rachitic pelvises, which he sold to Warren together with the skeleton. Warren Anatomical Museum Records, 1835–2010 (inclusive), 1971–1991 (bulk), Harvard Medical Library, Francis A. Countway Library of Medicine, Boston, MA.

5. J. Lawson Dick, *Rickets: A Study of Economic Conditions and Their Effects on the Health of the Nation, in Two Parts Combined in One Volume, Freely Illustrated* (New York: E. B. Treat, 1922); "Dick, John Lawson," *Plarr's Lives of the Fellows Online*, accessed November 1, 2023, https://livesonline.rcseng.ac.uk/client/en_GB/lives/search/results?qu=%22RCS: E003963%22&rt=false | | |IDENTIFIER| | |Resource+Identifier.

6. Dick, *Rickets*, 79.

7. Dick, *Rickets*, 85.

8. On Grafton Elliot Smith hauling a mummy in a Cairo taxi, see Arthur C. Aufderheide, *The Scientific Study of Mummies* (Cambridge: Cambridge University Press, 2003), 378. In 1896, within a year of Wilhelm Roentgen's discovery, Berlin physician W. Koenig x-rayed a mummified cat, and the eminent physician and public health activist Rudolf Virchow radiographed artifacts from Heinrich Schliemann's Troy excavations. A year later in London, Egyptologist W. M. Flinders Petrie x-rayed human mummies' bones to determine age at death; a Dr. Bloch claimed his place in the history of paleoradiology by being the first to x-ray a complete mummified body. Mario G. Fiori, Maria Grazia Nunzi, "The Earliest Documented Applications of X-rays to Examination of Mummified Remains and Archaeological Materials," *Journal of the Royal Society of Medicine* 88 (1995): 67–69, citing E. H. Ackerknecht, *Rudolf Virchow: Doctor, Statesman, Anthropologist* (Madison: University of Wisconsin Press, 1953).

9. Photographer Albert Londe x-rayed a mummy hand and forearm in 1897. Rethy K. Chhem and Don R. Brothwell, *Paleoradiology: Imaging Mummies and Fossils* (Berlin: Springer, 2008), 3; for Dr. Bloch and the first x-ray of a complete mummy body, see Françoise Dunand and Roger Luchtenberg, *Mummies and Death in Egypt*, trans. David Lorton (Ithaca, NY: Cornell University Press, 2006), 137.

10. Aufderheide says Smith examined ca. 30,000 mummies (obviously not full autopsies). Aufderheide, *Scientific Study of Mummies*, 13.

11. Dick, *Rickets*, 288.

12. Dick, *Rickets*, 16. But rickets is not completely absent in Egypt today, nor—recent archaeological finds suggest—in the distant past. I. A. Sabri, "A Consideration of the Possible Causes of the Prevalence of Rickets in Egypt," *Acta Paediatrica* 16 (1933): 442–46, argues that UV might be less available in urban Egypt than you might think and argues for widespread low-level rickets there. As for antiquity, a recent excavation in Egypt shows signs of rickets in poor ancient Egyptians. Traci Watson, "Ancient Egyptian Cemetery Holds Proof of Hard Labor," *National Geographic News*, March 14, 2013, https://www.nationalgeographic.com/history/article/130313-ancient-egypt-akhenaten-amarna-cemetery-archaeology-science-world.

13. Dick, *Rickets*, 288.

14. Soranus of Ephesus, *Gynaecia* II, 44 in Soranus, *Soranus' Gynecology*, trans. Owsei Temkin (Baltimore, MD: Johns Hopkins University Press, 1951), 116.

15. On significant examples from medieval England, see Ann Stirland, "The Human Bones," in Brian Ayres, ed., "Excavations within the North-East Bailey of Norwich Castle, 1979," *East Anglian Archaeology* 28 (1985); and Christopher Daniel, *Death and Burial in Medieval England 1066–1550* (New York: Routledge, 2005), 129–130. Of forty-nine infants exhumed from a twelfth-century cemetery in the Baltic port city of Schleswig, at the base of the Jutland Peninsula, four showed marked signs of rickets. G. Hühne-Osterloh and G. GrupeSource, "Causes of Infant Mortality in the Middle Ages Revealed by Chemical and Palaeopathological Analyses of Skeletal Remains," *Zeitschrift für Morphologie und Anthropologie* 77, no. 3 (April 1989): 247–58.

16. Traci Watson, "Skeletons Show Rickets Struck the Medici Family," *Nature*, June 7, 2013, http://www.nature.com/news/skeletons-show-rickets -struck-the-medici-family-1.13156; V. Giuffra, A. Vitiello, D. Caramella, A. Fornaciari, D. Giustini, and G. Fornaciari, "Rickets in a High Social Class of Renaissance Italy: The Medici Children," *International Journal of Osteoarchaeology* 25 (2015): 608–24.

17. "Archaeologists Discover London's Black Death Mass Grave," Medievalists.com, accessed August 6, 2021, https://www.medievalists .net/2014/03/archaeologists-discover-londons-black-death-mass-grave/. Carbon dating and other evidence showed that the bodies were interred during an epidemic of plague. Four of the skeletons (16 percent) bore clear signs of rickets. The nature of the burials suggests the victims were poor.

18. J. McSherry, "Failure to Thrive: Did Bonnie Prince Charlie Really Have Rickets?" *Journal of Medical Biography* 3, no. 2 (1995): 110–13.

19. J. J. Keevil, "The Illness of Charles, Duke of Albany (Charles I), From 1600 to 1612: An Historical Case of Rickets," *Journal of the History of Medicine and Allied Sciences* 9, no.4 (1954): 407–19; T. E. C., "Princess Elizabeth, Second Daughter of King Charles I, and Rickets," *Pediatrics* 64, no. 2 (1979): 241.

20. "Rickets" appeared earlier than 1634 in a handwritten receipt book and in a 1640 Botanical (J. Parkinson, *Theatrum Botanicum: The Theatre of Plants or an Herball of Large Extent* [London, 1640]). Both are cited in Jeffrey O'Riordan and Olav Bijvoet, "Rickets before the Discovery of Vitamin D," *BoneKEy Reports* 3 (2014): 1–6, https://doi.org/10.1038/bonekey.2013.212. See also Gill Newton, "Diagnosing Rickets in Early Modern England:

Statistical Evidence and Social Response," *Social History of Medicine* 35, no. 2 (October 5, 2021): 566–88, https://doi.org/10.1093/shm/hkab019.

21. This is especially true since the new social history's embrace of postmodern sociology and linguistics. The naming that is their subject carries far more weight than nominally joining disease with discoverer. As Phil Brown put it, "Our conception of medicalization . . . involves social control at very routine levels of socialization, labeling of behavior, and prescriptions for medical intervention. Diagnosis is central to such social control, since giving the name has often been the starting point for social labelers." Charles Rosenberg, "Disease in History: Frames and Framers," *Milbank Quarterly* 67 (1989): 1–15; and Phil Brown, "Naming and Framing: The Social Construction of Diagnosis and Illness," *Journal of Health and Social Behavior*, extra issue (1995): 34–52.

22. Peter M Dunn, "Francis Glisson (1597–1677) and the 'Discovery' of Rickets," *Archives of the Diseases of Children, Fetal and Neonatal Edition* 78 (1998): F154–F155. *De Rachidite* was something of a committee report, with fellow College of Physicians members George Bate and Ahasuerus Regemorter listed on the title page as assisting. J. L. H. O'Riordan, "Perspective: Rickets in the 17th Century," *Journal of Bone and Mineral Research* 21, no. 10 (2006): 1506–10. For the English translation, see Francis Glisson, George Bate, and Ahasuerus Regemorter, *A Treatise of the Rickets, Being a Disease Common to Children*, trans. Philip Armin (London: Cole, 1651).

23. Daniel Whistler, "De Morbo Puerili Anglorum, Quem Patrio Idiomate Indigenae Vocant the Rickets" (diss., Leyden, 1645). At least one other publication describing and naming "rickets" appeared before Glisson's: Arnold Boote's (1649) *Observationes Medicae de Affectibus Omissis* [*Medical Observations on Neglected Ailments*] (London: Thomas Newcomb, 1649). Boote's report remained obscure far longer than Whistler's, likely because the word "rickets" did not appear in the title.

24. Augusto A. Litonjua, *Vitamin D and the Lung: Mechanisms and Disease Associations* (New York: Springer, 2012): 5.

25. Samuel Gee, in a mid-nineteenth century overview, expressed his gratitude that Glisson's "rachitis" won out: "I need not say that the name Rachitis dates from Glisson; and we can hardly be surprised that it was quickly accepted, if only to escape from the heirs of the fertile invention of Dr. Whistler, his, 'paedocephalarthroncia,' 'paedosplanchnosteocace,' and other mighty inkhornisms." Samuel Gee, "On Rickets," *Saint Bartholomew's Hospital Reports* 4 (1868): 80; Denis Gibbs, "Rickets and the Crippled Child: An Historical Perspective," *Journal of the Royal Society of Medicine* 87 (December 1994): 729–32, quote 729.

26. Glisson did not long dwell on the recent etymology of the word: "The most received and ordinary Name of this Disease, is The Rickets: But who baptiz'd it, and upon what occasion or for what reason, or whether by chance or advice it was so named, is very uncertain." Simon Mays describes several instances of "rickets" being used prior to Glisson and Whistler. Simon Mays, "The Epidemiology of Rickets in the 17th–19th centuries: Some Contributions from Documentary Sources and Their Value to Palaeo-pathologists," *International Journal of Paleopatholology* 23 (December 2018): 88–95, https://doi.org/10.1016/j.ijpp.2017.10.011.

27. One almost laughable origin story comes from the biographer John Aubrey, from his 1656 history of his native Wiltshire: "About 1620 one Ricketts of Newberye, a Practitioner in Physick, was excellent at the Curing Children with swoln heads, and small legges: and the Disease being new, and without a name, he being famous for the cure of it, they called the Disease the Ricketts." Turning to Glisson and company, he observed, "And now 'tis good sport to see how they vex their Lexicons, and fetch it from the Greek [rachitis] the back bone." David Le Vay, "On the Derivation of the Name 'Rickets,'" *Proceedings of the Royal Society of Medicine* 68, no. 1 (January 1975): 46–50.

28. Le Vay, "Derivation of the Name." A small disclaimer is called for: I am not asserting that Glisson et al. intended a pun. I merely entertain the possibility, throwing one more small brand onto the bonfire of speculation about the etymology of "rickets."

29. Lawson Dick discusses a 1647 letter of Mary Verney, in England, writing to her husband, Sir Ralph Verney, who was in exile in France in the Royalist cause. Mary describes their children's legs: "Jack his legges are most miserable, crooked as every I saw any child's." Dick, *Rickets*, 311.

30. Glisson, Bate, and Regemorter, *Treatise*, 3.

31. Dick, *Rickets*, 278.

32. "The Thames Frost Fairs," *Historic UK*, accessed November 1, 2023, http://www.historic-uk.com/HistoryUK/HistoryofEngland/The-Thames -Frost-Fairs/. The freezing of the Thames is an inexact measure, it turns out, due to the confounding impact of changes in the design of the London Bridge and other factors that affected formation of ice dams on the river. See P. D. Jones, T. J. Osborn, and K. R. Briffa, "The Evolution of Climate over the Last Millennium," *Science*, New Series 292, no. 5517 (April 27, 2001): 662–67. Historical demographer John Kolmos argues that the climatic effects on grain production dramatically reduced the stature of the average Frenchman. John Komlos, "An Anthropometric History of Early-Modern France," *European Review of Economic History* 7, no. 2 (2003): 159–89. Wolfgang Behringer argues for the relationship between cold weather

and witchcraft mania in "Climatic Change and Witch-Hunting: The Impact of the Little Ice Age on Mentalities," in Christian Pfister, et al., eds., *Climatic Variability in Sixteenth-Century Europe and Its Social Dimension* (Heidelberg: Springer, 1999): 335–51.

33. Glisson, quoted in Dick, *Rickets*, 283.

34. Daniel Whistler, quoted by Jeffrey L. H. O'Riordan, "Reply: Rickets: Vitamin D and Calcium Deficiency," *Journal of Bone and Mineral Research* 22, no. 4 (2007): 639.

35. Science historian Antonio Clericuzio argues that "Glisson's changing views of spirits exemplify the transformation which occurred in his physiological ideas, namely, the abandonment off Galenic humoral medicine and the adoption of chemistry as the basis of physiology." Antonio Clericuzio, "The Internal Laboratory: The Chemical Reinterpretation of Medical Spirits in England (1650–1680)," in Piyo Rattansi, ed., *Alchemy and Chemistry in the 16th and 17th Centuries* (Dordrecht, Netherlands: Kluwer Academic Publishers, 1994): 51–83.

36. Sara Pennell, "Consumption and Consumerism in Early Modern England," *The Historical Journal* 42, no. 2 (June 1999): 549–64.

37. Kimberly Poitevin, "Inventing Whiteness: Cosmetics, Race, and Women in Early Modern England." *Journal for Early Modern Cultural Studies* 11, no. 1 (2011): 59–89.

38. Dick, *Rickets*, 310.

39. Baynard E. Floyer, *The History of Cold Bathing* (London: Smith and Walford, 1706), quoted in Gibbs, "Crippled Child," 730.

40. "Rickets: Vitamin D and Calcium," *Journal of Bone and Mineral Research* 22, no. 4 (2007): 638; Valerie Fildes, "The English Wet-Nurse and Her Role in Infant Care, 1538–1800," *Medical History* 32 (1988): 142–73.

41. Layinka M. Swinburne, "Rickets and the Fairfax Family Receipt Books," *Journal of the Royal Society of Medicine* 99, no. 8 (2006): 391–95, quote 392.

42. Kenneth Dewhurst, "Post-Mortem Examination on Case of Rickets Performed by John Locke," *British Medical Journal* 2 (5317) 1962: 1466. Locke found a cardiac malformation—a shunting between the right and left atria, which cardiologists recently argued was likely what is now known as an Ebstein's anomaly. A. N. Williams, N. Wilson, and R. Sunderland, "Philosopher, Pediatrician, Pathologist? John Locke's Thoughts on Rhicketts and a Missed Case of Ebstein's Anomaly," *Pediatric Cardiology* 30, no. 2 (July 30, 2008): 191–93; Rudolf Virchow, "Das Normale Knochenwachsthum und die Rachitische Störung Desselben," *Archiv für pathologische Anatomie und Physiologie und für klinische Medicin* 5 (1853): 409–507, https://doi.org/10.1007/BF01879060.

43. William David Osler, "Thesis on Rickets," University of Edinburgh, 1896.

44. Dick, *Rickets*, 288.

45. William Blake, "London," "The Chimney Sweeper," in *Songs of Innocence and of Experience* (London: Brimley Johnson, 1901).

46. William Blake, *Europe: A Prophecy* in *William Blake, Complete Writings*, ed. Geoffrey Keynes (Oxford: Oxford University Press, 1984), 243.

47. William Blake, *Milton, a Poem* (London: William Blake, 1808–1811), 2.

48. London's population growth exceeded 700 percent from 1662 to 1861; for 1662 John Graunt estimated the population at 384,000. W. G. Hoskins, *Local History in England* (London: Longman, 1984); for 1861 (3,188,485) see David T. Sunderland and Michael Ball, *An Economic History of London 1800–1914* (London: Routledge, 2001), 42. Coal consumption more than kept up with the rising population, doubling between 1830 and 1850; twenty-five years later, it had doubled again. See S. Mosley, *The Chimney of the World: A History of Smoke Pollution in Victorian and Edwardian Manchester* (Cambridge: White Horse, 2001), cited in Anne Hardy, "Commentary: Bread and Alum, Syphilis and Sunlight: Rickets in the Nineteenth Century," *International Journal of Epidemiology* 32 no. 3 (2003): 337–40.

49. The city's population went from 18,000 in 1750 to 338,000 in 1861.

50. Alexis de Tocqueville, "Journey to England (1835)," in *Journeys to England and Ireland*, trans. George Lawrence and K. P. Mayer, ed. J. P. Mayer (New Haven, CT: Yale University Press, 1958), 107.

51. Frederick Engels, *The Condition of the Working-Class in England in 1844, with Preface Written in 1892*, trans. Florence Kelley Wischnewetzky (London: Swan Sonnenshchein, 1892), 154.

52. Excerpt from "Promenades in London" (1840), in Flora Tristan, *Utopian Feminist: Her Travel Diaries and Personal Crusade*, ed., trans. Doris Beik and Paul Beik (Bloomington: Indiana University Press, 1993), 62.

53. Flora Tristan, *Promenades dans Londres*, 2nd ed. (Paris: Delloye, 1840), 96.

54. Flora Tristan, trans. Dennis Palmer and Giselle Pinceti, *Flora Tristan's London Journal, 1840* (Charlestown, MA: Charles River Books, 1980), 215. The passage in the original reads, "Quel air méphitique doit régner dans ces demeures! la figure des enfants en porte témoignage. – Rien de plus rachitique, de plus cadavéreux que ces petis ètres_—l'extrême maigreur, le teint blême, les yeux mornes, joints à l'excessive saleté et aux haillons qui les couvrent, offrent le spectacle le plus digne de compassion!" Tristan, *Promenades dans Londres*, 341–42.

55. This description draws on the Google Books Ngram for "rickety chair," "rickety table," "rickety house," "rickety child," 1800 to 2000, accessed

November 2, 2023, https://books.google.com/ngrams/graph?content
=rickety+chair%2Crickety+house%2Crickety+child&year_start=1800&year
_end=2000&corpus=en-2019&smoothing=3. Google's Ngram Viewer
displays the frequency that a given word or phrase appeared in the printed
sources published from 1500 to the present that have been scanned by the
Google Books project.

56. Charles Dickens, *A Tale of Two Cities: In Two Volumes, Volume 1*
(Leipzig: Tauchnitz, 1859), 88; Dawkins is introduced in chapter 8 of *Oliver
Twist* (London: Richard Bentley, 1839), 123. For the incidence of "rickets"
and related terms in Dickens's novels, I simply did full-text searches of the
texts available online at the Project Gutenberg site. Twice in the later novels, Dickens uses "knock-kneed" humorously: "whitewashed knock-knee'd
letters on a brew house sign" in *Great Expectations*, and in *Our Mutual
Friend* to disparage someone's intelligence: "It was constitutionally a knock-
knee'd mind and never very strong upon its legs."

57. For example, *Uncommercial Traveler* contains three references to
bowlegged individuals. Most gripping is a young lover he sees in a church-
yard, whose legs were "as much in the wrong as mere passive weakness of
character can render legs." Charles Dickens, *All the Year Round: A Weekly
Journal*, 9 (London: Chapman and Hall, 1863), 494.

58. James Greenwood, *The Wilds of London* (London: Chatto and
Windus, 1874): 143. "Little Bob in Hospital" appeared originally in the
Evening Star, February 25, 1867, 4, and was included in his collection of
essays published in 1874 as *The Wilds of London*. A generation after Little
Bob's first appearance, Greenwood republished the story yet again, giving
the little waif a more famous name: "Tiny Tim in Hospital," *Daily Telegraph*,
December 26, 1881.

59. Greenwood, *Wilds of London*, 148.

60. The Ormond Street Hospital was the first children's hospital in
Britain; its foundation in 1852 stands both as a high point in Victorian
efforts to help poor children and an enormous resource for research and
training in the then-new field of pediatric medicine. Charles Dickens had
been an important supporter of its foundation, speaking at fundraisers
and highlighting, in his fiction and journalism, the importance of such
institutions. Twice Greenwood reprinted his rescue story, partly to secure
public support for the hospital. See Katharina Boehm, "'A Place for More
Than the Healing of Bodily Sickness': Charles Dickens, the Social Mission
of Nineteenth-Century Pediatrics, and the Great Ormond Street Hospital
for Sick Children," *Victorian Review* 35, no. 1 (Spring 2009): 153–174, quote
154–55.

61. Greenwood, *Wilds of London*, 148.

62. Greenwood says that the children were "all glad to see the doctor—or magician, as I have called him." Greenwood, *Wilds of London*, 148.

63. Engels, *Condition*, 102. He did also implicate working conditions,
notably double shifts that kept juvenile mill workers at their posts through
the long night, without proper sleep in the day. Engels described one rickets
case "who got into this condition in Mr. Douglas' factory in Pendleton, an
establishment which enjoys an unenviable notoriety among the operatives
by reason of the former long working periods continued night after night."

64. A typical general medical text of the late Victorian era stated clearly
all three themes. "Rickets is one of the most common of diseases. . . .
Unfavorable circumstances, want of light and air, contribute largely to
its development. Hence the disease is more common among the poor
than among the well-to-do. It is especially abundant in great cities with a
large proletariat, in densely populated quarters, in the overcrowded, ill-
ventilated dwellings of the poor." Hugo Ziemssen, Albert Henry Buck, and
George Livingston Peabody, eds., *Cyclopædia of the Practice of Medicine*, 16
(New York: William Wood, 1877), 170–71.

65. "Geographical Distribution of Rickets, Acute and Subacute
Rheumatism, Chorea, Cancer, and Urinary Calculus in the British Islands,"
The British Medical Journal 1, no. 1464 (January 19, 1889): 113–16, quote
113.

66. One physician reported in 1867 an incidence of 30 percent among
the patients at the London Hospital for Sick Children. Gee, "On Rickets,"
69–80. J. Lawson Dick reported on an 1897 survey showing that nearly half
of children between six months and three years of age in Parisian hospitals
were rachitic; doctors on the continent reported rates as high as 80 percent,
in Prague and Bohemia, and over 50 percent in Frankfort. Dick, *Rickets*, 30.

67. W. Macewan, "Osteotomy with an Enquiry into the Aetiology
and. . . . ," London 1880, quoted in Gibbs, "Crippled Child," 729–32n15.

68. John Snow, "On the Adulteration of Bread as a Cause of Rickets,"
[first published in *Lancet* 1857; ii: 4–5]; repr. *International Journal of
Epidemiology* 32 (2003): 336–37, quote 336; Hardy, "Bread and Alum."

69. Hardy, "Bread and Alum."

70. Cheadle was president of the British Medical Association's Section
of Diseases of Children; W. B. Cheadle, "A Discussion of Rickets," *British
Medical Journal* 2, no. 1456 (November 24, 1888): 1145–52, quote 1145.

71. Greater consideration for skin color and latitude would follow. Until
well after the 1880s, when Theobald Palm published his findings on data he
collected from observers from around the world, there was no comparative
epidemiological explanation tying sunlight to the incidence of rickets, a
topic taken up in chapter 3.

72. Diagnosing Tiny Tim retrospectively has been a closet industry among historically inclined physicians for decades. Pediatric nephrologist Russell Chesney provides some historiographic perspective in his brief for Tim suffering from a combination of rickets and TB. Russell W. Chesney, "Environmental Factors in Tiny Tim's Near-Fatal Illness," *Archives of Pediatric Adolescent Medicine* 166, no. 3 (2012): 271–75, https://doi.org/10.1001/archpediatrics.2011.852.

Chapter 2

1. Guanahani is the Taino name for the Bahamian island. Columbus referred to it as such in early letters, though he later referred to the island as San Salvador. Questions about on which island Columbus first landed have driven a cottage industry of speculation. Douglas Peck, "Re-Thinking the Columbus Landfall Problem," *Terrae Incognitae* 28, no. 1 (January 1996): 12–35, https://doi.org/10.1179/tin.1996.28.1.12.

2. Columbus's manuscript *Diario* no longer exists; the historian and Dominican friar Bartolome de Las Casas abstracted it some forty years later. This translation is from Margarita Zamora, *Reading Columbus* (Berkeley: University of California Press, 1993).

3. For example, Cabeza de Vaca recorded this encounter: "All the Indians we had seen in Florida to this point were archers, and since they are so tall and they are naked, from a distance they look like giants. They are quite handsome, very lean, very strong and light-footed." Alvar Nunez Nunez and Cabeza de Vaca, "Account" (1541), *Early Visions Bucket* 8, accessed November 2, 2023, https://digitalcommons.usf.edu/early_visions_bucket/8.

4. Recently anthropologists examined the skeletons of hundreds of sailors recovered from the wreck of their sixteenth-century British ship, the *Mary Rose*. Many of the skeletons exhibited clear evidence of severe childhood rickets. Jemma G. Kerns, Kevin Buckley, Anthony W. Parker, Helen L. Birch, Pavel Matousek, Alex Hildred, and Allen E. Goodship, "The Use of Laser Spectroscopy to Investigate Bone Disease in King Henry VIII's Sailors," *Journal of Archaeological Science* 53 (January 1, 2015): 516–20, https://doi.org/10.1016/j.jas.2014.11.013.

5. To be specific, had the sailors been extremely vitamin D deficient, they would be at risk of osteomalacia, so-called adult rickets.

6. James Lind, *Treatise of the Scurvy. In Three Parts* (Edinburgh: Sands, Murray and Cochran for A Kincaid and A Donaldson, 1753). Arun Bhatt, "Evolution of Clinical Research: A History Before and Beyond James Lind," *Perspectives in Clinical Research* 1, no. 1 (2010): 6–10, PMCID: PMC3149409.

7. Analysis of the bones of settlers from Columbus's second voyage suggest that scurvy might have killed many men who persisted in eating only the familiar foods they had brought from Spain, eschewing "exotic" local fruits and vegetables whose vitamin C content would have saved them. V. Tiesler, A. Coppa, P. Zabala, and A. Cucina, "Scurvy-Related Morbidity and Death among Christopher Columbus' Crew at La Isabela, the First European Town in the New World (1494–1498): An Assessment of the Skeletal and Historical Information," *International Journal of Osteoarchaeology* 26, no. 2 (2016), https://doi.org/10.1002/oa.2406.

8. Today's medical historians are accustomed to Charles Rosenberg's general guidance for framing disease to reveal the many ways in which health and sickness are social constructions. Those trained in or toiling in social or cultural history are long accustomed to thinking of the idea of race as just that: an idea—a social construction erected to guide the exercise of power by those on top. We tend to discount explanations that emphasize racial or ethnic differences, arguing that while the data, even where they can be trusted, demonstrate association, they fail to establish a causal relationship. For years, historians have had worldly and wise guides through these troubled shoals, guides such as Kenneth Kiple and Todd L. Savitt, Keith Wailoo, Sam Roberts, and Susan Reverby, and others to whom I am deeply indebted.

9. With apologies to William Cronon, who concluded his famous article about narrative: "I would urge upon environmental historians the task of telling not just stories about nature, but stories about stories about nature." "A Place for Stories: Nature, History, and Narrative," *Journal of American History* 78, no. 4 (1992): 1375.

10. Though they may try to conceal them: John Murchison placed an ad in 1831 for a fugitive from slavery named Simon, who was "very knock kneed" and "remarkably artful." Murchison noted that Simon "may endeavor to hide his being knock-kneed by wearing large pantaloons." *Carolina Observer* (Fayetteville, NC), February 10, 1831, 3, Digital Library on American Slavery/North Carolina Runaway Slave Notices, https://dlas .uncg.edu/notices/notice/197/.

11. Stephens's ad appeared in the *Carolina Sentinel* on April 28, 1826, Digital Library on American Slavery/North Carolina Runaway Slave Notices, https://dlas.uncg.edu/notices/notice/1704/.

12. Or only a month later. "Run away . . . in April last," refers to either six months before the ad was placed, or eighteen months. If eighteen, Hagai ran off within a month of her indenture. The basic facts about Sexton appear in a number of books. The additional facts in my account come from Paul Heinegg, "Free African Americans in Colonial Virginia, North Carolina,

South Carolina, Maryland and Delaware," accessed January 20, 2024, http://www.freeafricanamericans.com/Scott-Skipper.htm.

13. Virginia Gazette (Dixon & Hunter), Williamsburg, December 26, 1777. "TWENTY DOLLARS REWARD. FOR apprehending and delivering to the subscriber, in Fredericksburg, a mulatto wench named SALLY . . . A reward of 20 dollars will also be given for apprehending and delivering to me, a mulatto servant named HANKEY (alias Hagai Sexton) who went off about 4 years ago, and has been living these last two years in Albemarle county, where she passed for a free woman, and is still there, protected and harboured by a parcel of free Mulattoes. WILLIAM SMITH." Accessed November 2, 2023, http://www2.vcdh.virginia.edu/gos/search/relatedAd.php?adFile=vg1777.xml&adId=v1777122319.

14. On the subject of disabilities in general in the context of American slavery, see Dea Hadley Boster, *African American Slavery and Disability: Bodies, Property and Power in the Antebellum South, 1800–1860* (New York: Routledge, 2012).

15. *The Pennsylvania Chronicle, and Universal Advertiser*, July 10, 1769, 199.

16. The ad for Enoch Calvert appeared in the *Richmond Enquirer*, April 18, 1823, 4; only a few years earlier, ads for John Craig, wanted in the "Atrocious Murder" of a justice of the peace in Pennsylvania described him as "stoop shouldered, stout built, a little knock kneed, very much sun burnt." In August of 1817, newspapers posted the offer of a $300 award for the capture of one Craig, a blacksmith, *National Aegis* (Worcester, MA) August 6, 1817, 2; Craig allegedly murdered the justice, Edward Hunter, to keep Hunter from testifying about a contested will the justice had written. John Woolf Jordan, *A History of Delaware County, Pennsylvania, and its People*, vol. 1 (New York: Lewis Historical, 1914), 261.

17. In 1974, as historians were starting to apply quantitative methods to the analysis of runaway ads, Peter Wood referred to *his* set of ads thus: "The runaways found in this survey number well over three hundred, yet this group must represent little more than the top of an ill-defined iceberg." His caveat would apply many times over in trying to use the ads to derive a solid assessment of the distribution of rickets. Peter Wood, *Black Majority: Negroes in Colonial South Carolina from 1670 through the Stono Rebellion* (New York: Knopf, 1974), 240.

18. "Burial Vaults Inspire a Celebration of a Church Opposed to Slavery," *New York Times*, October 8, 2014; Meredith Ellis, "The Children of Spring Street: Rickets in an Early Nineteenth-Century Congregation," *Northeast Historical Archaeology* 39, no. 1 (2010): 120–33.

19. "Skeleton Crew—Building at Trump Tower Halted after Bones Found," *New York Post*, December 13, 2006, 13.

20. Douglas B. Mooney, Edward Morin, Robert Wiencek, and Rebecca White, "Archaeological Investigations of the Spring Street Presbyterian Church Cemetery, 244–246 Spring Street," URS Corporation, December 2008, http://s-media.nyc.gov/agencies/lpc/arch_reports/1138.pdf.

21. Dunlap, "Burial Vaults," A 33.

22. Meredith Ellis, *The Children of Spring Street: The Bioarchaeology of Childhood in a 19th Century Abolitionist Congregation* (Cham, Switzerland: Springer, 2019), 40; preliminary findings appeared in Ellis, "Children of Spring Street,"120–33. Ellis focused on long bones, measuring degrees of curvature, and noting other visible signs, as well as making microscopic observations of subtle metabolic effects of rickets and other dietary deficiencies.

23. Ellis, "Children of Spring Street," 129.

24. Michael L. Blakey, "Introduction," *The New York African Burial Ground: Unearthing the African Presence in Colonial New York; Vol 1: The Skeletal Biology of the New York African Burial Ground*, ed. Michael L. Blakey and Lesley M Rankin-Hill (Washington, DC: Howard University Press, 2009), 3.

25. Blakey, *New York African Burial*, 7.

26. Of the 285 skeletons with "observable lower limb bones," 11.9 percent had bilaterally bowed legs (they counted only "individuals who expressed bowing bilaterally"); adults' rate was 14.4 percent; "subadults'" rate was 6.5 percent. Blakey, *New York African Burial*, 194–95.

27. The children of the African Burial Ground exhibited bowlegs at 6.5 percent, compared with 52.78 precent in the Spring Street collection. Ellis, *Children of Spring Street*, 40.

28. "*Presumably* never buried," because it is not clear whether some of these skeletons had been provided by "Resurrection Men"—gravediggers who disinterred recently buried corpses and sold them to anatomists or medical schools. Michael Sappol, *A Traffic of Dead Bodies: Anatomy and Embodied Social Identity in Nineteenth Century America* (Princeton, NJ: Princeton University Press, 2001).

29. Carlina De la Cova, "Race, Health, and Disease in 19th Century-Born Males," *American Journal of Physical Anthropology* 144 (2011): 526–37.

30. J. Park West, "Rickets: Its Prevalence in Eastern Ohio," *University Medical Magazine* 8, no. 1 (1895): 30–33.

31. Leon Mead, *The Bowlegged Ghost and Other Stories: A Book of Humorous Sketches, Verses, Dialogues and Facetious Paragraphs* (New York: Werner, 1899), 17–25.

32. Only one of the wandering dead is described as other than white: Mead describes the parade of second-class ghosts as including "divers other ghostly freaks and monstrosities, including a two-headed colored ghost,

concerning whose earthly identity I failed to learn—making in all seventy-five grisly spectres" (22).

33. *Augusta Herald*, August 6, 1800, 4; this ad also appeared in "Trivia," *William and Mary Quarterly* 6, no. 2 (April 1949): 285–89 (citing as *Virginia Argus*, October 31, 1800).

34. "An Editor Lost," *Macon Telegraph*, May 16, 1886, 3. The insult made its way into political discourse too, as when the *Idaho Statesman* reflected, in 1868, on the political climate in the neighboring state and "congratulated" "those knock-kneed weak-backed Republicans in the last Oregon legislature, who were too conservative to vote for a registry law." "The Election in Oregon," *Idaho Statesman*, April 9, 1868, 2.

35. Elbert Hubbard, *No Enemy but Himself*, repr. in *Selected Writings of Elbert Hubbard: His Mintage of Wisdom, Coined from a Life of Love, Laughter and Work*, vol. 11 (New York: W. H. Wise, 1922), 107.

36. Alex E. Sweet, "The Bill Snort Letters," *Columbus* (Georgia) *Daily Enquirer*, July 5, 1891, 3. A popular joke highlighted the issue for women trading form-concealing dresses for revealing beachwear: "That girl must be bow-legged," said the clerk. "Why?" asked the floorwalker. "She wanted a bathing suit with an extra long skirt." "Jumping at Conclusions" [quoting the *Chicago Record-Herald*], *The Technical World Magazine* 6–7 (1906): 447.

37. Howard Fielding [pseudonym for Charles W. Hooke, though article byline says "Copyright 1892 by James W. Johnson"], "Did Anybody See Her? Howard Fielding Vainly Searches for a Mysterious Woman," *Salt Lake Weekly Tribune*, September 8, 1892, 7.

38. Fielding, "Did Anybody," 7.

39. "People in Tights," *Columbus* (Georgia) *Daily Enquirer*, February 12, 1886, 7.

40. Gordon Samples, *Lust for Fame: The Stage Career of John Wilkes Booth* (Jefferson, NC: McFarland, 1982), 121; "in face and figure," Arthur F. Loux, *John Wilkes Booth: Day by Day* (Jefferson, NC: McFarland, 2014), 5.

41. Michael W. Kauffman, *American Brutus: John Wilkes Booth and the Lincoln Conspiracies* (New York: Random House, 2007), 118.

42. Samples, *Lust for Fame*, 121.

43. "Colts Colts Colts! They Pranced Around in Their Own Pasture Like Kentucky 3-Year-Olds," *Fort Worth Register*, April 27, 1897, 5.

44. The Dallas paper introduced new player, Charles Krehmeyer, as "very bowlegged, a left-hand thrower and a hitter for your life." *Dallas Morning News*, January 20, 1890, 5; *Colliers* reminisced in 1908 about Frank Chance's rise a decade earlier: "The real beginning of the Chicago world's championship team was in March, 1898, when a big, bow-legged, rather awkward young player came from the Pacific Coast to be tried as a catcher."

Hugh Stuart Fullerton, "The Making of a World's Championship Ball Club," *Colliers*, August 8, 1908, 118. Gene Rye's brief fame is described in Mark McGuire and Michael Sean Gormley, *Moments in the Sun: Baseball's Briefly Famous* (Jefferson, NC: McFarland, 1999), 82–85.

45. *New York American* (November 19, 1907) quoted at http://www
.baseball-almanac.com/quotes/quowagn.shtml; " 'Bugs' Raymond Holds Pirates Safe: National League Leaders Get One Run from Eight Widely Scattered Hits. Loving Cup for Wagner: Famous Player Receives Evans Trophy for Leading the Major Leagues in Batting Last Year," *New York Times*, May 21, 1909, 7; Gomez quote, "Game's Greatest Shortstop Honus Wagner Dead at 81," *Spencer* [Iowa] *Daily Reporter*, December 6, 1955. Wagner's accomplishments as a player are well recorded. Historians of public health praise him for a choice he made in 1909, a decision that resulted in the 1909 Honus Wagner baseball card becoming the most valuable in history, even more highly prized than Babe Ruth's rookie card. Baseball cards in 1909 were produced by the American Tobacco Company and distributed in cigarette packs. A nonsmoker himself, and concerned that his card's inclusion might "influence children to purchase tobacco products," Wagner revoked his contract with the tobacco company—and got his cards pulled. Fewer than two hundred Honus Wagner cards were distributed, making them one of the most sought-after trading cards. In 2022 one sold for $7.25 million. Robert Proctor, *Golden Holocaust: Origins of the Cigarette Catastrophe and the Case for Abolition* (Berkeley: University of California Press, 2011): 88; Michael Shapiro, "Honus Wagner T-206 Card Sells for Record $7.25 Million," August 4, 2022, https://www.si.com/mlb/2022/08
/04/honus-wagner-t206-card-sells-record-price.

46. Ogden Nash, "Line-Up for Yesterday: An ABC of Baseball Immortals," *Sport*, January 1949, 32.

47. San Jose, CA *Evening News*, August 30, 1895, 1.

48. [William Alden], "The Coming Leg," *New York Times*, January 22, 1880, 4.

49. The 1905 exhibition became notorious when arch obscenity fighter Anthony Comstock confiscated five hundred pounds of promotional posters ("vile handbills," he called them) and arrested MacFadden. *Physical Culture* 10, no. 6 (December 1903): 547; Dave Tell, *Confessional Crises and Cultural Politics in Twentieth-Century America* (University Park: Pennsylvania State University Press, 2012): 23.

50. "So-Called Beauty Contest: 'Physical Culture Show' Begins in Madison Square Garden," *New York Times*, October 2, 1905, 5.

51. Joe Beadis, "Honus Wagner Puts Away Spike, Uniform for Keeps," *St Petersburg Times*, February 17, 1952, 17.

52. Henry James Finn, ed., *American Comic Annual* (Boston: Richardson, Lord & Holbrook, 1831),138.

53. [James Henderson?], "Perault; or Slaves and Their Masters," *Tait's Edinburg Magazine*, February 1843, 91.

54. For a full discussion of the relationship between disability and slavery—the "dual stigma of race and disability," see Boster, *African American Slavery*.

55. Mark Twain, *Mark Twain: The Complete Interviews*, ed. Gary Scharnhorst (Tuscaloosa: University of Alabama Press, 2006), 457. In another version of the story, the man never gave chase himself but "came after us with dogs." In this version, the boys plot a deadly revenge. Twain recalled to his boyhood friend that "we made up our minds we'd catch that n_____ and drown him." Thinking back on that revenge fantasy, Twain added: "John, if we had killed that man we'd have had a dead n_____ on our hands without a cent to pay for him." Albert Bigelow Paine, *Mark Twain, a Biography: The Personal and Literary Life of Samuel Langhorne Clemens*, vol. 3 (New York: Harper & Brothers, 1912), 1170.

56. Cynthy speaks in the exaggerated black "dialect" so common in white writers' prose of that era. Julia B. Tenney, "Mammy's Love-Story," *The Chautauquan* 33 (September 1901): 597–601. Sometimes the devaluation is clear even if the race of the sufferer is not. A story from late nineteenth-century Kansas purports to be mocking the federal government's growing reliance on bureaucracy and qualifications for civil service. It is not clear that this particular story is about an African American—dialect is ambiguous, and it clearly is not about a slave. Still, the same dynamic is at work: "'Well, sir,' said the old farmer, 'this here durned red-tape gover'ment is the devil. Why, you've got ter stan' a reg'lar school examination fer ever'thing! Fust, they turned John down fer the postoffice jest kaze he didn't know nothin' 'bout spellin' an' 'rithmetic, an' now they won't take him in the army kaze he's bow-legged in one leg an' knock-kneed in the other! How kin they expect people to live happy under a gover'ment like that!" "Starbeams," (Freemen) *Kansas City Star*, May 22, 1898.

57. Hinton Rowan Helper, *Nojoque: A Question for a Continent* (New York: Carleton, 1867), 69.

58. Louis Maurer, "An Heir to the Throne, or the Next Republican Candidate," New York, Currier & Ives, 1860, Library of Congress, accessed January 21, 2024, https://www.loc.gov/pictures/resource/pga.09074/. The cartoon is based on a Currier & Ives circus poster for P. T. Barnum's 1860 "What Is It?" exhibition; see CUNY Graduate Center, "The What Is It? Exhibit, The Lost Museum Archive," accessed November 3, 2023, https://lostmuseum.cuny.edu/archive/exhibit/what/.

59. Edgar Allan Poe, "The Journal of Julius Rodman, Chapter V," *Burton's Gentleman's Magazine*, May 1840, 206–10; for another example, a Vicksburg slaveholder in 1845 described an enslaved groom as "an excellent specimen of the old Virginia negro—very bow legged with a remarkably tall head with a crape on it." "Matters and Things at Vicksburg," *Spirit of the Times; A Chronicle of the Turf, Agriculture, Field Sports, Literature and the Stage*, April 26, 1845, 1.

60. The play in question is Edward Sheldon, *The N_____* (New York: Macmillan, 1915), 4; Joel Chandler Harris, "Death and the Negro Man," *Uncle Remus and His Friends* (1892), 36.

61. H. Seely Totten, "Dipping," *Godey's Lady's Book*, vol. 31, July–December, 1845, 199–201.

62. L. Beauregard Pendleton, "Plantation Fun and Folklore," *The Current*, April 19, 1884, 283–4.

63. Thomas Dixon, *The Clansman: An Historical Romance of the Ku Klux Klan* (New York: Doubleday, 1906): 241–43. D. W. Griffith's racist master-piece *Birth of a Nation*, based on Dixon's novel, depicted Aleck, but only in passing. Melvyn Stokes, *D. W. Griffith's* The Birth of a Nation*: A History of the Most Controversial Motion Picture of All Time* (New York: Oxford University Press, 2008): 197.

64. Robert Stewart, comment recorded in [Cincinnati] "Academy of Medicine Meeting of March 9th, 1886," *Cincinnati Lancet-Clinic*, vol. 55, 1886, 368–69. Forty years later, *Hygeia* still agreed, offering the observation "that negroes living in the temperate zones are likely to develop rickets." Olive Swanson, "Is Rickets a Normal Condition?" *Hygeia*, September 1928, 494–96, quote 496. Actually, the assumption goes back to colonial times. Caribbean plantation physician Robert Thomas dismissed rickets as a concern in his health manual for sugar planters: "This disease, although very frequently met with amongst children in cold climates, is almost wholly unknown to the inhabitants of warm ones, and need not therefore be particularly described." Robert Thomas, "Of the Rickets," *Medical Advice to the Inhabitants of Warm Climates, on the Domestic Treatment of all the Diseases Incidental Therein: With a Few Useful Hints to New Settlers, for the Preservation of Health, and the Prevention of Sickness* (London: J. Johnson, St. Paul's Church-Yard; J. Strahan, Strand; and W. Richardson, Royal-Exchange, 1790), 333.

65. Slavery's apologists certainly sought to make the comparison between healthy southern plantations and the deadly conditions of European factory cities. In 1822, Charleston attorney Edwin Clifford Holland published a proslavery pamphlet arguing that the dramatic natural increase of the African American population was "proof of the humanity with

which these people are treated." Holland asserted that "food is both more wholesome and more abundant that that of the laboring classes in other countries," with the result that "dropsies, rickets, scrofula, typhus fever, and the long train of disease which attend upon want and poverty, are far less frequent amongst our slaves, than in England, Scotland and Ireland." Edwin Clifford Holland, *A Refutation of the Calumnies Circulated against the Southern and Western States, Respecting the Institution and Existence of Slavery among Them* (Charleston, SC: A. E. Miller, 1822), 55.

66. Shane White and Graham White, "Slave Clothing and African-American Culture in the Eighteenth and Nineteenth Centuries," *Past & Present* 148 (August 1995): 149–86, 173; in 1938, former slave Frank Fikes recalled, "I went in my shirt tail until I was eleven or twelve years old. Back in slavery time boys did not wear britches." Works Progress Administration, *Slave Narratives: A Folk History of Slavery in the United States from Interviews with Former Slaves*, vol. 1, Arkansas Narratives (Washington, DC: Federal Writer's Project, United States Work Projects Administration, 1941), 285. The practice of swaddling offers a potential counterargument, but evidence points toward swaddling having played at best a small role in African American childrearing practices. Joseph E. Illick, "Childhood in Three Cultures in Early America," *Pennsylvania History: A Journal of Mid-Atlantic Studies* 64 (Summer 1997), 308–23.

67. Frederick Law Olmsted, *A Journey in the Seaboard Slave States; with Remarks on Their Economy* (New York: Dix and Edwards, 1856), 391; in 1937, ninety-eight-year-old Mattie Curtis told a WPA interviewer she went completely naked till she was fourteen. "I [was] naked like [that] when my nature come to me." Works Progress Administration, *Slave Narratives: A Folk History of Slavery in the United States from Interviews with Former Slaves*, vol. 11, North Carolina Narratives (Washington, DC: Federal Writer's Project, United States Work Projects Administration), 102.

68. Richard H. Steckel, "A Peculiar Population: The Nutrition, Health, and Mortality of American Slaves from Childhood to Maturity," *Journal of Economic History* 46, no. 3 (1986): 721–41; John M. Pettifor, "Nutritional Rickets: Deficiency of Vitamin D, Calcium, or Both?" *American Journal of Clinical Nutrition* 80, no. 6 (2004): 1725S–29; Ana L. Creo, Tom D. Thacher, John M. Pettifor, Mark A. Strand, and Philip R. Fischer, "Nutritional Rickets around the World: An Update," *Paediatrics and International Child Health* 37, no. 2 (2017): 84–98.

69. Kenneth Kiple, "Slave Child Mortality: Some Nutritional Answers to a Perennial Puzzle," *Journal of Social History* 10 (March 1977): 284–309.

70. Steckel, "Peculiar Population": "fell among or below," 727; European comparisons, 728.

71. Holland, *Refutation*, 55.

72. Perhaps the most notorious of these tracts comparing Southern slavery with Northern "Industrial Slavery" was George Fitzhugh's *Cannibals All! Or Slaves without Masters* (Richmond, VA.: A. Morris, 1857).

73. Jim Downs, *Sick from Freedom: African-American Illness and Suffering during the Civil War and Reconstruction* (New York: Oxford University Press, 2012); "attributable solely," *Richmond Dispatch*, November 10, 1863, quoted in Downs, *Sick from Freedom*, 40n107; "Poor dusky children," 23; see also Emily Spivack, "The Story of Elizabeth Keckley, Former-Slave-Turned-Mrs. Lincoln's Dressmaker," Smithsonian.com, April 24, 2013, https://www .smithsonianmag.com/arts-culture/the-story-of-elizabeth-keckley-former -slave-turned-mrs-lincolns-dressmaker-41112782/.

74. Scott Alan Carson, "The Effect of Geography and Vitamin D on African American Stature in the Nineteenth Century: Evidence from Prison Record," *Journal of Economic History* 68, no. 3 (September 2008): 812–31. This slight increase in stature for African Americans directly contrasted with a decline in stature for antebellum Americans as a whole—the so-called "Antebellum Puzzle." Carson finds fairly uniform height increases in the antebellum years, but a more complex story after the Civil War, with rural blacks doing better in terms of height (Carson uses height as proxy for vitamin D status).

75. "Have We Any Rickets among Us?" *Louisville Medical News* 8, no. 4 (July 26, 1879): 42–43.

76. Charles S. Shaw, "Rickets," *Medical and Surgical Reporter*, August 31, 1895, 258.

77. George N. Acker, "Rickets in Negroes" *Transactions of the American Pediatric Society* 6 (1894): 137–142. Acker's interpretation of the fallout of Reconstruction would soon find academic expression in the so-called Dunning School.

78. "Have We any Rickets?," 42–43.

79. Marion Post Wolcott, *Negro Children and Old Home on Badly Eroded Land Near Wadesboro, North Carolina, December 1938*, Library of Congress Prints and Photographs Division Washington, http://www.loc.gov/pictures /item/fsa2000031123/PP/. Wolcott was employed by the Farm Security Administration. But as if to reinforce the reality that rickets is no simple "race-specific disease," see the similar photo Arthur Rothstein shot, also for the Farm Security Administration, "Sharecropper's Child Suffering from Rickets and Malnutrition, Wilson Cotton Plantation, Mississippi County, Arkansas," August 1935, LC-USF33- 002002-M2 [P&P] | LC-USF33-002002-M2, https://www.loc.gov/pictures/item/2017720685/. For his 1992 biography of Wolcott, journalist Paul Hendrickson tracked down

the two children in the photo. The little boy, in his fifties then, still lived in Wadesboro. His bowlegs had straightened as he grew. His parents took him to a clinic where, he recalled, "the doctors wanted to break both my legs with mallets and reset them." But they brought him home and, he told Hendrickson, "wrapped me, and gave me some special exercising" and proved "all those doctors at the clinic wrong." Paul Hendrickson, *Looking for the Light: The Hidden Life and Art of Marion Post Wolcott* (New York: Knopf, 1992), 98.

80. Jerome C. Rose, "Biological Consequences of Segregation and Economic Deprivation: A Post-Slavery Population from Southwest Arkansas," *Journal of Economic History* 49, no. 2 (June 1989): 351–60.

81. Mitchell found that 49.8 percent of white children and 87.6 of Blacks showed "definite clinical signs of rickets." F. Thomas Mitchell, "Incidence of Rickets in the South," *Southern Medical Journal* 23, no. 3 (March 1930): 228–35.

Chapter 3

1. Correspondence from Joseph Stokes to Edwards A. Park, 22 March 1934, folder "Park, Edwards A #4," Joseph Stokes Papers B: St65P, American Philosophical Society, Philadelphia.

2. Correspondence from Dorothy Whipple to Harry L. Russell, 25 November 1933, folder "Milk, Wisconsin Fund," Joseph Stokes Papers B: St65P, American Philosophical Society, Philadelphia; "minimizing . . . sunlight," Milton Rapaport, Joseph Stokes, and Dorothy V. Whipple, "The Antirachitic Value of Irradiated Evaporated Milk in Infants," *Journal of Pediatrics* 6, no. 6 (June 1935): 799–808, quote 800.

3. Rapaport, Stokes, and Whipple, "Antirachitic Value," 799–808.

4. Arwady, "Uses of Rickets," 6.

5. Glisson, Bate, and Regemorter, *Treatise*, 159.

6. Glisson, Bate, and Regemorter, *Treatise*, 208–9.

7. William E. Horner, *The Home Book of Health and Medicine: Being a Popular Treatise* . . . (Philadelphia: Key and Biddle, 1834), 424.

8. "Society Reports: Clinical Society of Maryland," *Maryland Medical Journal* 31, no. 16 (August 4, 1894): 305–6.

9. "White plague" is attributed to Oliver Wendell Holmes, from an 1869 lecture. Allan B. Weisse, *The Staff and the Serpent: Pertinent and Impertinent Observations on the World of Medicine* (Carbondale: Southern Illinois University Press, 1998): 43. In 1908, William Osler famously misremembered a line from seventeenth-century poet John Bunyan, who referred to "Captain Consumption, with all his men of death" when Osler promoted pneumonia

to the status of "captain of the men of death." Michael Fry, *Landmark Experiments in Molecular Biology* (Cambridge, MA: Academic Press, 2016), 43.

10. Robert J. Lee, "On the Precedent Cause of Rickets," *Lancet* 2 (December 15, 1888): 1169–70 Abraham Jacobi's textbook *Diseases of Children* correctly asserted, "The fact that rachitis may often develop intrauterine is not equivalent with heredity," and provided several more or less anecdotal recent studies that attempted, unsuccessfully, to prove that rickets was inherited. He did not, however, think the question settled: "At all events *the heredity of rickets, the congenital occurrence, and the racial immunity* unite in a triad of facts which, if their entity could be established, would be of the greatest significance in the pathogenesis of this affection." Abraham Jacobi, ed. *Diseases of Children* (New York: D. Appleton, 1910): 245.

11. And while she admitted that "the children of the more robust mothers are not all knock-kneed and puny," they are, "in every instance, inferior—if not always physically, certainly in mental quality or in human charm—to those of the more womanly type." Arabella Kenealy, "Woman as an Athlete: A Rejoinder," *The Nineteenth Century* 45 (June 1899): 915–29, quotes 924–25.

12. Irving M. Snow, "An Explanation of the Great Frequency of Rickets among Neapolitan Children in American Cities," *Medical News*, September 22, 1894, 316–20; "still rare," 316; "first generation," 319.

13. His mixed success is dryly captured in the entry for him in a contemporary memoir and survey of missionaries to Japan: "Dr. Theobald Palm (Edinburgh Medical Mission), a man of rare gifts and spirit, spent several years at Niigata. During one year he treated 2,950 patients, this medical work being more than self-supporting. As a result of his labors and those of his Japanese associates, preaching was maintained for a time in thirteen different places, and eighty-eight were baptized. Dr. Palm's recall prevented the full ripening of results." M. L. Gordon, *An American Missionary in Japan* (Boston: Houghton Mifflin, 1893): 171.

14. A. Hamish Ion, *The Cross and the Rising Sun: The British Protestant Missionary Movement in Japan, Korea and Taiwan, 1865–1945* (Waterloo, Ontario: Wilfrid Laurier University Press), 45.

15. Palm, "The Geographic Distribution and Etiology of Rickets" *The Practitioner* 45 (1890): 273.

16. Palm, "Geographic Distribution," 328.

17. Palm, "Geographic Distribution," 333–34.

18. Palm, "Geographic Distribution," 335.

19. Palm, "Geographic Distribution," 336.

20. On Herschel, see Jack R. White, "Herschel and the Puzzle of Infrared," *American Scientist* 100, no. 3 (May–June 2012): 218–25, https://

doi.org/10.1511/2012.96.218; on Ritter's discovery and its reception, Jan Frercks, Heiko Weber, and Gerhard Wiesenfeldt, "Reception and Discovery: The Nature of Johann Wilhelm Ritter's Invisible Rays," *Studies in History and Philosophy of Science Part A* 40, no. 2 (June 2009):143–56. The literature on Röntgen's discovery is huge; for a brief overview, see John Maddox, "The Sensational Discovery of X-Rays," *Nature* 375 (May 18, 1995): 183.

21. Freund, *American Sunshine*; Elena Conis, "The Rise and Fall of Sunlight Therapy," *Los Angeles Times*, May 28, 2007, https://www.latimes.com /archives/la-xpm-2007-may-28-he-esoterica28-story.html.

22. Alan I. Marcus, ed., *Science as Service: Establishing and Reformulating American Land-Grant Universities, 1865–1930* (Birmingham: University of Alabama Press, 2015).

23. Christian Warren, *Brush with Death: A Social History of Lead Poisoning* (Baltimore: Johns Hopkins University Press, 2000), 36.

24. The best known of these hagiographies being, of course, Paul DeKruif, *Microbe Hunters* (New York: Harcourt, Brace, 1926). DeKruif also published a collection of essays on the heroes of nutrition and agricultural science, *Hunger Fighters* (New York: Harcourt, Brace, 1928).

25. The most cited study is by J. Koch (not, as some have mistakenly concluded, Robert Koch). J. Koch, "Untersuchungen über die Lokalisation der Bakterien, das Verhalten des Knochenmarkes und die Veränderungen der Knochen, insbesondere der Epiphysen, bei Infektionskrankheiten. Mit Bemerkungen zur Theorie der Rachitis," *Zeitschrift für Hygiene und Infektionskrankheiten* 1911; 69: 436–62. This is discussed in Fritz B. Talbot, "Recent Advances in Our Knowledge of Rachitis," *AJDC* 3 (1912): 112–23.

26. M. Allison Arwady examines this notion of "negative causality" in "Uses of Rickets," 15.

27. Alan M. Kraut. *Goldberger's War: The Life and Work of a Public Health Crusader* (New York: Hill and Wang, 2003); Sarah Laskow, "How Sick Chickens and Rice Led Scientists to Vitamin B1," *The Atlantic*, October 30, 2014.

28. Lederer, *Subjected to Science*.

29. Taking up some middle ground in this ethical abyss was the antivivisectionists' nightmare—the tradition described by science populizer Paul DeKruif as the relentless massacre of the guinea pigs in the name of scientific progress. From De Kruif's *Microbe Hunters*, chapter 6, "Roux and Behring: Massacre the Guinea Pigs."

30. Dating the "experiment" is a challenge. Some reports put it in 1889. But the *British Medical Journal* published commentary on it late in 1888, describing the subject lion cubs as now being eighteen months old, which puts the experiment back at least to early 1887. Bland-Sutton's January study,

published in January 1888 dates the cure earlier, perhaps 1885 or 1886. W. B. Cheadle, "A Discussion on Rickets, in the Section of Diseases of Children at the Annual Meeting of the British Medical Association, Held in Glasgow, August 1888," *British Medical Journal* 2, no. 1456 (November 24, 1888): 1145–52; J. Bland-Sutton, "Rickets in Monkeys, Lions, Bears and Birds," *Journal of Comparative Medicine and Surgery* 10, no. 1 (January 1888): 1–29.

31. "Bland-Sutton, Sir John (1855–1936)," *Plarr's Lives of the Fellows*, accessed November 2, 2023, https://livesonline.rcseng.ac.uk/client/en_GB /lives/search/results?qu=Bland-Sutton&te=; for details on the London School of Anatomy and Physiology, see "Cooke, Thomas, 1841–1899" *Plarr's Lives of the Fellows*, accessed November 2, 2023, https://livesonline.rcseng .ac.uk/client/en_GB/lives/search/results?qu=cook%2C+thomas&h=1.

32. Bland-Sutton, "Rickets in Monkeys." Note, he does discuss treatment of lions: "It may here be mentioned that some rickety cubs, which early manifested signs of rickets, were promptly fed on bone-dust and cod-liver oil, made a good recovery, and were alive and active, presenting no signs of paralysis two years afterwards" (25). It is not clear how Bland-Sutton was brought in on the case of the rickety lion cubs. The many subsequent published accounts always cast it in the passive voice (e.g., "was consulted about").

33. Russell W. Chesney and Gail Hedberg, "Metabolic Bone Disease in Lion Cubs at the London Zoo I 1889: The Original Animal Model of Rickets," *Journal of Biomedical Science* 17, no. S1 (2010): S36, http://www .jbiomedsci.com/content/17/S1/S36. Again, since Bland-Sutton did not specify, there is speculation as to the specific diet he suggested.

34. W. B. Cheadle, president of the British Medical Association's section on pediatrics, reporting Bland-Sutton's description of the experiment; Cheadle, "Discussion on Rickets," 1146.

35. "all signs," Cheadle, "Discussion on Rickets," 1146; Elmer Verner McCollum, *A History of Nutrition: The Sequence of Ideas in Nutrition Investigations* (Boston: Houghton Mifflin, 1957): 271–72. Bland-Sutton contrasted the unfavorable conditions in London's zoo with those in Dublin, Manchester, and some other British towns, where "lions can be reared successfully in captivity." Bland-Sutton, "Rickets in Monkeys," 26. This essay contains detailed drawings and descriptions of the rickety cubs, and briefly mentions curing them with "bone-dust and cod-liver oil" (p. 25).

36. Bland-Sutton, "Rickets in Monkeys."

37. Alfred F. Hess, "Newer Aspects of the Rickets Problem," *JAMA* 78, no. 16 (April 22, 1922): 1177–83.

38. Harry G. Day, "Elmer Verner McCollum," *Biographical Memoirs of the National Academy of Sciences* 45 (1974): 263–335 (275).

39. Both teams' work appeared in the same issue of the *Journal of Biological Chemistry* in 1913, but since McCollum and Davis submitted their manuscript a few weeks earlier, they have received credit for the discovery of what would eventually be labeled "Fat Soluble A" and eventually "Vitamin A." E. V. McCollum and M. Davis, "The Necessity of Certain Lipins in the Diet during Growth," *Journal of Biological Chemistry* 15 (1913): 167–75; the Yale team's findings: Thomas Osborne and Lafayette Mendel, "The Relation of Growth to the Chemical Constituents of the Diet," *Journal of Biological Chemistry* 15 (1913): 311–26.

40. Christopher Maternowsky, "Lady Mellanby's Dental Utopia," Nursing Clio, May 31, 2017, https://nursingclio.org/2017/05/31/lady-mellanbys-dental-utopia/.

41. This choice of subject is somewhat odd given that Mellanby's main research advisor in college was research committee member Frederick Garland Hopkins: famous for discovering tryptophan, who had coined the term "accessory food factors," and who certainly knew that McCollum and Davis had just published on discovering the fat-soluble antirachitic. J. Parascandola and A. J. Ihde, "Edward Mellanby and the Antirachitic Factor," *Bulletin of the History of Medicine* 51 (1977): 507–15, quote 509.

42. Henry H. Dale, "Edward Mellanby. 1884–1955," *Biographical Memoirs of Fellows of the Royal Society* 1 (November 1955): 192–222, 201; May Mellanby conducted the radiological studies at her institution; she turned her research material over to her husband. She achieved fame with her studies linking vitamin D and dental issues. On her dental studies, see Maternowsky, "Lady Mellanby's."

43. Edward Mellanby to Henry Dale, April 8, 1946, quoted in Parascandola and Ihde, "Edward Mellanby," 512.

44. Edward Mellanby, "Discussion on the Importance of Accessory Food Factors (Vitamines) in the Feeding of Infants," *Proceedings of the Royal Society of Medicine* 13 (1920): 57–98; quote 97.

45. Barbara J. Hawgood, "Sir Edward Mellanby (1884–1955) GBE KCB FRCP FRS: Nutrition Scientist and Medical Research Mandarin," *Journal of Medical Biography* 18, no. 3 (2010):150–57, https://doi.org/10.1258/jmb.2010.010020.

46. Edward Mellanby to Henry Dale, April 8, 1946, quoted in Parascandola and Ihde, "Edward Mellanby," 514.

47. Edward Mellanby, "The Part Played by an Accessory Factor in the Production of Experimental Rickets," *Proceedings of the Physical Society* 52 (January 26, 1918), xi–xii; E. Mellanby, "An Experimental Investigation on Rickets," *Lancet* 1 (1919): 407–12; quoted in Kumaravel Rajakumar, "Vitamin D, Cod-Liver Oil, Sunlight, and Rickets: A Historical Perspective,"

Pediatrics 112, no. 2 (August 1, 2003): e132–35, https://doi.org/10.1542 /peds.112.2.e132, e133.

48. E. V. McCollum, N. Simmonds, J. E. Becker, and P. G. Shipley, "An Experimental Demonstration of the Existence of a Vitamin which Promotes Calcium Deposition," *Journal of Biological Chemistry* 53 (1922): 293–98.

49. Dick, *Rickets*, 85.

50. Dick, *Rickets*, 83.

51. Leonard Findlay, "Diet as a Factor in the Cause of Rickets," *Archives of Pediatrics* 38 (1921): 151–62, quote 151; the "tradition" privileging dietary cause was evident, according to Findlay, in how often Bland-Sutton's experience with those lions in the London Zoo was mentioned. Findlay corresponded with Bland-Sutton, whom he claimed later backed off of diet as cause for the lions' rickets, preferring a microbial cause.

52. Mellanby, "Discussion on the Importance," 95.

53. Findlay, "Diet as a Factor," 155.

54. Russell W. Chesney provides a brief introduction to these studies in "Theobald Palm and his Remarkable Observation: How the Sunshine Vitamin Came to be Recognized," *Nutrients* 4, no. 1 (2012): 42–51; A. F. Hess and M. Weinstock, "Anti-Rachitic Properties Imparted to Inert Fluids and to Green Vegetables by Ultra-Violet Irradiation," *Journal of Biological Chemistry* 62 (1924): 301–13; H. Steenbock and A. Black, "Fat Soluble Vitamins XVII: The Induction of Growth-Promoting and Calcifying Properties in a Ration by Exposure to Ultraviolet Light" *Journal of Biological Chemistry* 61 (1924): 405–22.

55. Rima D. Apple, "Patenting University Research: Harry Steenbock and the Wisconsin Alumni Research Foundation," *Isis* 80, no. 3 (September 1989): 374–95, 374.

56. "The Cause of Rickets," *Nature* 110 (1922): 137–38.

57. Findlay himself wrote essays from the perspective of "The Glasgow School"—for example, the very essay *Nature* was reviewing: Leonard Findlay, "A Review of the Work Done by the Glasgow School on the Aetiology of Rickets," *Lancet* 199, no. 5148 (1922): 825–31.

58. "Discussion on the Etiology of Rickets," *British Medical Journal* 2 (November 4, 1922): 846–58, quote 850.

59. Palm, "Geographic Distribution."

60. H. S. Hutchison and S. J. Shah, "The Aetiology of Rickets, Early and Late," *Quarterly Journal of Medicine*, 15 (1922): 167–94; see also Bivins, "'English Disease.'"

61. Leonard Findlay, "The Etiology of Rickets: A Clinical and Experimental Study," *British Medical Journal* 2 no. 2479 (July 4, 1908): 13–17, quotes 15.

62. D. Noel Paton, Leonard Findlay, and Alexander Watson, "Observations on the Cause of Rickets," *British Medical Journal* 2 (December 7, 1918): 625–26.

63. H. Steenbock and E. B. Hart, "The Influence of Function on the Lime Requirements of Animals," *Journal of Biological Chemistry* 14 (1913): 59–73. This article barely links sunlight and calcium, but in 1921, Hart and Steenbock clarified: "Marked differences in the amount of calcium eliminated in the feces of a milking goat when that animal was changed from old dried roughage to green pasture. The goat in question was confined throughout, only its food was changed. Evidently something had been ingested with the green material that allowed a more perfect skeletal storage of calcium." E. B. Hart, H. Steenbock, and C. A. Hoppert, "Dietary Factors Influencing Calcium Assimilation," *Journal of Biological Chemistry* 47, no. 1 (1921): 33–49. A misguided interpretation of the 1913 study occasionally pops up in professional and popular accounts. Led astray by Steenbock's expression "green pasture," writers assume the healthy goats had been put out in green pasture. For example: "In 1913, University of Wisconsin's H. Steenbock and E. B. Hart [showed] that lactating goats kept indoors lose a great deal of their skeletal calcium, whereas those kept outdoors do not." "Unraveling the Enigma of Vitamin D," in *Beyond Discovery: The Path from Research to Human Benefit* (Washington: National Academy of Sciences; 2000), http://www.nasonline.org/publications/beyond-discovery/vitamin-d.pdf.

Chapter 4

1. R. A. Hobday, "Sunlight Therapy and Solar Architecture," *Medical History* 42 (1997): 455–72; Kumaravel Rajakumar, Susan L Greenspan, Stephen B. Thomas, and Michael F. Holick, "Solar Ultraviolet Radiation and Vitamin D: A Historical Perspective," *American Journal of Public Health* 97, no. 10 (October 2007): 1746–54, 1749, https://doi.org/10.2105/AJPH.2006.091736.

2. Kurt Huldschinsky, "Heilung von Rachitis durch Kunstliche Hohensonne," [Rickets Cured by Ultraviolet Irradiation]" *Deutstsche Medizinische Wochenschrift* 45 (1919): 712–13, cited in Rajakumar, Greenspan, Thomas, and Holick, "Solar Ultraviolet Radiation," 1748.

3. Kenneth J. Carpenter, "Harriette Chick and the Problem of Rickets," *Journal of Nutrition* 138, no. 5 (May 2008): 827–32; H. Chick, E. J. Palzell, and E. M. Hume, *Studies of Rickets in Vienna 1919–1922*, Medical Research Council, Special Report: No. 77, 1923.

4. Alfred F. Hess and Lester J. Unger, "An Interpretation of the Seasonal Variation of Rickets," *AJDC* 22 (1921): 186–92, quote 186.

5. Hess lived in New York for most of his life. Born in 1875, the son of publisher Selmar Hess, he received a Harvard undergraduate degree and earned his medical degree from Columbia University's College of Physicians and Surgeons; from 1915 to 1931, he was clinical professor of pediatrics at New York University. "Dr Alfred F. Hess Dead at Age of 58," *New York Times*, December 7, 1933, 23.

6. Alfred F. Hess and Lester J. Unger, "Prophylactic Therapy for Rickets in a Negro Community," *JAMA* 69, no. 19 (November 10, 1917): 1583–86, quote 1583; see Kumaravel Rajakumar and Stephen B. Tomas, "Reemerging Nutritional Rickets: A Historical Perspective," *Archives of Pediatric and Adolescent Medicine* 159 (2005): 335–41.

7. Hess and Unger, "Prophylactic Therapy," 1585. In the end, significant attrition left only forty-nine infants who received regular cod liver oil, and sixteen "controls"—children who received regular visits and exams but whose mothers chose not to administer cod liver oil. In the discussion following Hess's reading of the paper before the AMA's Section on Preventive Medicine and Public Health, Haven Emerson, New York City's Commissioner of Health dismissed Hess's call for the Health Department to distribute cod liver oil free of charge at baby stations. But he celebrated the educational value of the study, telling an anecdote Hess had left out of his remarks: "The mothers are now coming to his clinic saying: 'I want to have cod liver oil for my baby.' He asks them, 'Why do you want it?' They have little swellings around the end of the long bones; they have got little lumps on their wrists.' The mothers are beginning to realize what is happening to the babies" (1586).

8. Alfred F. Hess and Lester J. Unger, "The Clinical Role of the Fat-Soluble Vitamin: Its Relation to Rickets," *JAMA* 74, no. 4 (January 24, 1920): 217–23. Hess put it a bit more directly several years earlier, in the context of a trial of pertussis vaccines: "These are some of the conditions which are insisted on in considering the course of experimental infection among laboratory animals, but which can rarely be controlled in a study of infection in man." Alfred F. Hess, "The Use of a Series of Vaccines in the Prophylaxis and Treatment of an Epidemic of Pertussis," *JAMA* 63, no. 12 (1914): 1007–11, quote 1007.

9. Alfred F. Hess, "Institutions as Foster Mothers for Infants," *Archives of Pediatrics* 33 (1916): 96–106.

10. Alfred F. Hess, "The Use of a Simple Duodenal Catheter in the Diagnosis and Treatment of Certain Cases of Vomiting in Infants," *AJDC* 3, no. 3 (1912): 133–52; A. F. Hess and Mildred Fish, "Infantile Scurvy: The Blood, the Blood Vessels, and the Diet," *AJDC* 8 (1914): 386–405. See Susan E. Lederer and Michael A. Grodin, "Historical Overview: Pediatric

Experimentation," in *Children as Research Subjects: Science, Ethics, and Law*, eds. Michael A. Grodin and Leonard H. Glantz (New York: Oxford University Press, 1994), 6.

11. Hess and Unger, "Clinical Role."

12. Konrad Bercovici, "Orphans as Guinea Pigs," *The Nation*, June 29, 1921, 911–13. Bercovici wanted to be clear he was calling for medicine to self-regulate, lest "anti-vivisectionist fanatics and the various freaks and cranks who are constitutionally hostile to organized medicine and its progress" seize the wheels and force government to intervene legislatively, "a development which is neither desirable nor necessary" (913). Susan Lederer puts this moment into its full context within a long debate over vivisection and child experimentation in her *Subjected to Science*, esp. 106–7.

13. "Amazing Growth of a Philanthropy," *New York Times*, July 18, 1921, 22.

14. "Orphans and Dietetics," *American Medicine* 27 (August 1921): 394–95; "owned and controlled," *American Medicine* 1 no. 1.

15. Hess's experiments with lab animals led to the discovery (nearly simultaneously with Steenbock's nearly identical, but patent-winning discovery) of how to enrich foods using UV light and his research supporting Otto Windaus's Nobel Prize winning research on the composition of vitamin D.

16. William M. Schmidt, "Some Kind of Social Doctor: Martha May Eliot, 1891–1978," *Pediatrics* 63, no. 1 (January 1979): 146–49. Bert Hansen, "Public Careers and Private Sexuality: Some Gay and Lesbian Lives in the History of Medicine and Public Health," *American Journal of Public Health* 92, no. 1 (2002): 36–44, https://doi.org/10.2105/ajph.92.1.36.

17. Martha M. Eliot, "The Control of Rickets: Preliminary Discussion of the Demonstration in New Haven," *JAMA* 85, no. 9 (August 29, 1925): 656–63, quote 657.

18. "Control groups" in Eliot's study did not mean what it usually connotes in a clinical trial; instead the study counted as controls infants "born outside the demonstration district" but concurrent with the study's subjects, who were not given cod liver oil or instructed in giving sun baths. The other controls were older infants "who had had no antirachitic treatment." Eliot, "Control of Rickets," 660–61.

19. Eliot, "Control of Rickets," 661.

20. Martha M. Eliot, *The Effect of Tropical Sunlight on the Development of Bones of Children in Puerto Rico*, US Dept. Labor, Children's Bureau Publication No. 217 (Washington, DC: United States Government Printing Office, 1933). The study included 584 children. On clinical exam 50 received a positive diagnosis (8 percent): 46 were "of slight degree," 3 of moderate degree, and 1 of marked degree." The one "marked" case is described at

page 120. Eliot responded to an earlier study finding high rates of rickets in Puerto Rico—Eliot clearly disproved this, explaining that Puerto Rican physicians reported malnutrition deaths as "rachitismo," similar to the US practice of catchall phrases like "marasmus."

21. Viosterol is the brand name assigned to the form of vitamin D produced by irradiating foodstuffs containing ergosterol. Although supplement advertisers consistently capitalized the "V," most references to irradiated ergosterol dropped the capitalization, and Viosterol became viosterol much in the way Bayer's Aspirin quickly became plain aspirin. Here I will use the uncapitalized form unless quoting ads or referring directly to Viosterol as a patented product.

22. Rapaport, Stokes, and Whipple, "Antirachitic Value," 799.

23. Apple, "Patenting University Research," 386. For an exhaustive and critical examination of Steenbock, WARF, and the patenting of vitamin D fortification, see Walters, "Before the Foundation."

24. Marietta Eichelberger, "The History of Evaporated Milk," *Journal of Home Economics* 24, no. 9 (September 1932): 793–97; in 1910, American per-capita consumption stood at 5.5 pounds per year; in 1935 it had climbed to 16.2 pounds, a 195 percent increase. Bureau of the Census, *Historical Statistics of the United States, Colonial Times to 1970*, Part 1, G 906, 331.

25. Janet Golden, *Babies Made Us Modern: How Infants Brought America into the Twentieth Century* (Camden, NJ: Rutgers University Press, 2018), 106.

26. W. T. Nardin, vice president of Pet Milk Company, to John Claxton Gittings, CHOP, February 23, 1934, Joseph Stokes Papers, American Philosophical Society B: St65P/ Folder "Milk-Wisconsin Fund #4."

27. The previous study was undertaken at St. Vincent's Hospital, a large foundling home. John Mitchell, John Eiman, Dorothy Whipple, and Joseph Stokes, "Protective Value for Infants of Various Types of Vitamin D Fortified Milk: A Preliminary Report," *AJPH* 22, no. 12: 1220–29.

28. Gittings to Harry L. Russell (WARF) December 7, 1933; it is cold comfort that at the very least, these children, unlike those in the St. Vincent's experiments, were not orphans; on the other hand, it is only too easy to imagine the shoddy nature of informed parental consent in effect in early Depression-era Philadelphia.

29. According to data supplied by the Franklin Institute, the low temperature on January 2, 1934, was twenty-four degrees Fahrenheit, with a high of thirty-nine degrees. Although the institute recorded no precipitation that day, their data do not speak to whether skies were cloudy or clear. "Historical Weather Data for Philadelphia," Franklin Institute, accessed

November 2, 2023, https://www.fi.edu/en/science-and-education/collection
/historical-weather-data-philadelphia.

30. Rapaport, Stokes, and Whipple, "Antirachitic Value," 800.

31. Rapaport, Stokes, and Whipple, "Antirachitic Value," 808.

32. The tested milk contained 125 USP units of vitamin D per pint can (14.5 ounces); Rapaport, Stokes, and Whipple, "Antirachitic Value," 808.

33. Form letter from the Carnation Company, 8 June 1934, folder "Milk, Wisconsin Fund F #3," Joseph Stokes Papers, American Philosophical Society B: St65P/.

34. Russell, director of WARF, to Stokes, 5 June 1934, "Milk, Wisconsin Fund F #3."

35. Marietta Eichelberger, director, Nutrition Services, Irradiated Evaporated Milk Institute (Chicago), to Dr. J. C. Gittings, Children's Hospital of Philadelphia, 25 October 1934, "Milk, Wisconsin Fund F #2."

36. Milton Rapoport and Joseph Stokes, "The Antirachitic Value of Irradiated Evaporated Milk and Irradiated Whole Fluid Milk in Infants," *Journal of Pediatrics* 8, no. 2 (1936): 154–60.

37. Rhoads et al., "Studies on the Growth and Development of Male Children Receiving Evaporated Milk," *Journal of Pediatrics* 26, no. 5 (1945): 415–54.

38. AMA Committee on Foods, "Vitamin D Milk," [nd (1935?)], ms Stokes papers, folder "American Medical Association Committee on Foods: Vitamin D Milk."

39. Rena Crawford and G. Richarda Williamson, "A Study of the Incidence of Rickets in a Group of Infants in New Orleans and an Effort to Prevent the Disease in a Certain Number of that Group," *New Orleans Medical and Surgical Journal* 83, no. 4 (October 1930), 219–27. For historical context regarding experiments on children, see Cynthia A. Connolly, *Children and Drug Safety: Balancing Risk and Protection in Twentieth-Century America* (New Brunswick, NJ: Rutgers, 2018).

40. John W. M. Bunker and Robert S. Harris, "A Reappraisal of Vitamin D Milks," *NEJM* 219, no. 1 (1938): 9–12.

41. The Wisconsin Alumni Research Foundation certainly seemed to, and expressed the opposite of discomfiture about these experiments, bragging in a flyer, "Brief Excerpts from Scientific Literature," about the lengths to which Rapaport and Stokes went to "minimize any antirachitic effect" of sunlight in the darkened ward where their human subjects contracted rickets. Wisconsin Alumni Research Foundation, "Brief Excerpts from Scientific Literature," 1937, pp. 17–18, accessed January 22, 2024, https://www.warf.org/wp-content/uploads/2020/08/Timeline-Sunshine-15.pdf.

42. Children's Hospital of Philadelphia does not seem to have been a major center for orthopedic surgery in the 1930s; statistics from annual reports from the period show only a couple of corrective surgeries for conditions likely associated with rickets. Children's Hospital of Philadelphia, Annual Reports, 1931–1933, Historical Medical Library, College of Physicians.

43. At the start of the twentieth century, a New York orthopedic surgeon reported that in the previous ten years, patients with rachitic deformities made up 15 percent of the forty-two thousand patients treated at the Hospital for the Ruptured and Crippled. "about one-half came on account of bow-legs, over one-quarter for knock knees," Henry Ling Taylor, "The Surgery of Rickets," *JAMA* 39, no. 15 (October 11, 1902): 901–3.

44. Roscoe C. Giles, "Rickets: The Surgical Treatment of the Chronic Deformities of, with Emphasis on Bow-Legs and Knock Knees," *Journal of the National Medical Association* 14, no. 1 (1922): 9–11.

45. C. D. T. James, "Sir William Macewen," *Proceedings of the Royal Society of Medicine* 67 (1974): 237–42. For details on Macewan's osteotomy, see Peltier, *Orthopedics*, 95.

46. William Macewen, *Osteotomy: With an Inquiry into the Aetiology and Pathology of Knock-Knee, Bow-Leg, and Other Osseous Deformities of the Lower Limbs* (London: J&A Churchill, 1880), 161.

47. Macewen, *Osteotomy*, 161–168.

48. T. Prowse, S. Saunders, C. Fitzgerald, L. Bondioli, and R. Macchiarelli, "Growth, Morbidity, and Mortality in Antiquity: A Case Study from Imperial Rome," in Tina Moffat and Tracy Prowse, eds. *Human Diet and Nutrition in Biocultural Perspective: Past Meets Present* (New York: Berghahn Books, 2010): 173–198, 187.

49. E. Muirhead Little, "Glisson as an Orthopaedic Surgeon," *Journal of the Royal Society of Medicine* 19 (May 1926): 111–22.

50. Andry's tract appeared in French in 1741, with an English edition published two years later. Nicolas Andry, *Orthopaedia: The Art of Correcting and Preventing Deformities in Children* (London: A. Millar, 1743), vol. 1, 282.

51. Snow, "Explanation," 319.

52. David B. Levine, "Hospital for Special Surgery: Origin and Early History; First Site, 1863–1870," *Hospital for Special Surgery Journal* 1 (2005): 3–8; "light gymnastics," Newton Shaffer, "On the Care of Crippled and Deformed Children," *The New York Medical Journal*, July 9, 1898, 37–40.

53. David B. Levine, "The Hospital for the Ruptured and Crippled: Knight to Gibney, 1870–1887," *Hospital for Special Surgery Journal* 2 (2006): 1–6.

54. David B. Levine, "The Hospital for the Ruptured and Crippled, Entering the Twentieth Century, ca. 1900 to 1912," *Hospital for Special Surgery Journal* 3 (2007): 2–12.

Chapter 5

1. The image's visual approach is striking, employing three very different styles for the three main elements: the crawling baby is drawn somewhat realistically, while the rickety shadow is almost mannerist in its exaggerated distortion; the fish seem to have swum in from a low-budget animated cartoon. "Supplements the Sun . . . Removes the Shadow of Rickets," Mead Johnson & Company, Mead's Oleum Percomorphum, *Journal of the Florida Medical Association* 36 (1949–50), back cover.

2. As will be explained below, Meads proudly advertised only to physicians, hewing to a principle they and other pharmaceutical companies referred to as "Ethical Marketing."

3. Freund, *American Sunshine*.

4. To be specific, as will be important later, viosterol was synthesized by irradiating ergosterol, found in plants, producing vitamin D2; vitamin D3, on the other hand, is the vitamin produced in animals or synthesized by irradiating animal cholesterols. For statistics on cod liver consumption, see USDA Economic Research Service, "US Fats and Oils," Statistical Bulletin No. 376, 1966.

5. My respect for the important role codfish played in this history will not extend as far as the ill-prepared Harvard history student who, confronted with an exam prompt to discuss the Newfoundland fisheries dispute of 1890, began, "This topic has been exhaustively examined from the viewpoint of the United States, Great Britain, and Canada. I shall discuss it from the viewpoint of the fish." Morton Keller, *The Unbearable Heaviness of Governing: The Obama Administration in Historical Perspective* (Stanford, CA: Hoover, 2013), 1.

6. *Gadus* comprises the three most important cod (Atlantic, Pacific, and Greenland) as well as the Alaska pollock. On the many species of fish *called* cod, see Mark Kurlansky, *Cod: A Biography of the Fish That Changed the World* (New York: Walker, 1997; repr., New York: Penguin Books, 1998), 37–38. On the subject of what vitamin D is to the cod that its livers contain so much, see E. -J. Lock, R. Waagbø, S. Wendelaar Bonga, and G. Flik. "The Significance of Vitamin D for Fish: A Review." *Aquaculture Nutrition* 16, no. 1 (2010): 100–16, https://doi.org/10.1111/j.1365-2095.2009.00722.x.

7. Arthur Spiess, Kristin Sobolik, Diana Crader, John Mosher, and Deborah Wilson, "Cod, Clams and Deer: The Food Remains from

Indiantown Island," *Archaeology of Eastern North America* 34 (2006): 141–
87. Anne Karin Hufthammer, Hans Høie, Arild Folkvord, Audrey J. Geffen,
Carin Andersson, and Ulysses S. Ninnemann, "Seasonality of Human
Site Occupation Based on Stable Oxygen Isotope Ratios of Cod Otoliths,"
Journal of Archaeological Science 37 (2010): 78–83; Wim Van Neer, Anton
Ervynck, Loes J. Bolle, Richard S. Millner, and Adriaan D. Rijnsdorp, "Fish
Otoliths and Their Relevance to Archaeology: An Analysis of Medieval, Post-
Medieval, and Recent Material of Plaice, Cod and Haddock from the North
Sea," *Environmental Archaeology* 7 (2002): 61–76.

8. The finest grades of salt cod went to European markets, but fisheries sold the worst cuts, soon known as West Indies cod, to Caribbean slave
labor camps. Kurlansky, *Cod*, 80–82.

9. Esben M. Olsen, Mikko Heino, George R. Lilly, M. Joanne Morgan,
John Brattey, Bruno Ernande, and Ulf Dieckmann, "Maturation Trends
Indicative of Rapid Evolution Preceded the Collapse of Northern Cod,"
Nature 428 (April 29, 2004): 932–35.

10. Food and Agriculture Organization of the United Nations, "The Cod,"
accessed October 20, 2023, http://www.fao.org/3/x5911e/x5911e01.htm.

11. "heaping together," Thomas Percival, "Observations on the
Medicinal Uses of the Oleum Jecoris Aselli, or Cod Liver Oil, in the Chronic
Rheumatism, and other Painful Disorders" *London Clinical Medical Journal*
3 (1783): 392–401, quote 395; Ludovicus Josephus De Jongh, *The Three
Kinds of Cod Liver Oil: Comparatively Considered with Reference to Their
Chemical and Therapeutic Properties*, transl. from German by Edward Carey
(Philadelphia: Lea & Blanchard, 1849), 420.

12. "How They Make Cod Liver Oil," *Scientific American* 4, no. 24 (June 15,
1861): 379.

13. "Cod Liver Oil." *Columbus* [Georgia] *Tri-Weekly Enquirer*, Decem-
ber 10, 1857, 2; H. Fineberg and A. G. Johanson, "Industrial Use of Fish
Oils," US Department of the Interior, Fish and Wildlife Service Bureau of
Commercial Fisheries, m 230-232 GPO 929-311(1967?), 228–29; Dale
Gilbert Jarvis and Terra M. Barrett, "The Historical Use of Ochre Pigments
in Newfoundland and Labrador," Heritage NL: Fieldnotes Series, 003,
March 2019, https://historicpaintnl.ca/use-of-ochre-pigments/.

14. "Reviews and Bibliographical Notices," *Dublin Quarterly Journal of
Medical Science* 9, no. 18 (1850): 412–52, quote 413; a New England physician concurred, noting that "among the fishermen of our own Continent,
it has long been highly valued for its curative power in rheumatism and its
analogous diseases." Daniel M'Ruer, "Cod Liver Oil," *Boston Medical and
Surgical Journal* 40 (March 21, 1849): 137–40, quote 137, https://doi.org
/10.1056/NEJM184903210400702.

15. Percival, "Observations," 396.

16. M'Ruer, "Cod Liver Oil," 137.

17. The quote is from a 1779 report on fish oil, not cod liver oil specifically: *Medical and Philosophical Commentaries by a Society of Physicians in Edinburgh* 6, pt.1 (1779), 95, reprinted in Thomas E. Cone Jr., "The First Report of the Value of Fish Liver Oil as a Cure for Rickets," *Pediatrics* 40, no. 2 (1967): 183.

18. Bacteriologist Ruth Guy summarized the process in 1923: "In each country the experience was the same: cod liver oil was used by the fishing people and peasantry; then accidentally observed by some physician, tried by him, and so made known generally to the profession." Ruth A. Guy, "The History of Cod Liver Oil as a Remedy," *American Journal of the Diseases of Children* 26, no. 2 (1923): 115.

19. M'Ruer, "Cod Liver Oil," 137; Percival, "Observations," 396.

20. Guy, "History," 112.

21. R. S. Trotter, "The Introduction of Cod-Liver Oil into Medical Practice in This Country," *The British Medical Journal* 2, no. 2270 (July 2, 1904): 44.

22. "Cod-Liver Oil," *Scientific American* 24, no 7 (February 11, 1871): 102.

23. Thomas Percival, *Medical Ethics; or, a Code of Institutes and Precepts, Adapted to the Professional Conduct of Physicians and Surgeons* (Manchester: S. Russell, 1803); Sara Patuzzo, Giada Goracci, and Rosagemma Ciliberti, "Thomas Percival. Discussing the Foundation of Medical Ethics," *Acta Biomedica* 89, no. 3 (2018): 343–48.

24. John V. Pickstone and Stella V. F. Butler, "The Politics of Medicine in Manchester, 1788–1792: "Hospital Reform and Public Health Services in the Early Industrial City," *Medical History* 28, no. 3 (1984): 227–49, quote 230, https://doi.org/10.1017/s0025727300035924.

25. Percival opened his report by humorously acknowledging that his readers would suspect he was contributing to "the progressive accumulation . . . of inactive remedies," but assured that introducing cod liver oil would "be doing some service to the healing art." Percival, "Observations," 393; on Darbey, 397.

26. Guy, "History," 113.

27. Percival, "Observations," 400.

28. Darbey, quoted in Percival, "Observations," 401; Darbey had been the Manchester Infirmary's resident apothecary since 1765 and house surgeon since a few years later. He held the position until 1787. Edward Mansfield Brockbank, *Sketches of the Lives and Work of the Honorary Medical*

Staff of the Manchester Infirmary (Manchester: University Press, 1904): 108–9.

29. Dr. Bradshaw, "On the Use of Cod-Liver Oil in Chronic Rheumatism," *Provincial Medical & Surgical Journal* 9, no. 53 (December 31, 1845): 753–57.

30. "Mr. Wesener upon the Good Effects of Cod-Liver Oil in Rheumatism and Gout," *Providence Patriot*, August 6, 1825, 1; "An Obstinate Remedy," *New Orleans Times* (Louisiana), February 26, 1867, 10. A similar but dark and humorless story hinged on a German immigrant mishearing his doctor's recommendation as "dog liver oil," "A New Remedy," *Texas State Gazette* (Austin), June 6, 1857, 1.

31. *Medical and Philosophical Commentaries by a Society of Physicians in Edinburgh* 6, pt. 1 (1779), 95. Found in Cone, "First Report," 183.

32. Guy, "History," 114, quoting Armand Trousseau.

33. John Hughes Bennett, *Treatise on the Oleum Jecoris Asselli, or Cod Liver Oil, as a Therapeutic Agent in Certain Forms of Gout, Rheumatism, and Scrofula: With Cases* (London: Highley, 1841), "various states" vii; "complete cure . . . excite attention," 101.

34. De Jongh, *Three Kinds*, 114.

35. Robert Hunter, "The Treatment of Consumption by Allopathy, Homeopathy, Water Cure and Medicated Inhalation Practically Reviewed," *New York Herald*, April 29, 1857, 11.

36. On the inconsistency and potency factor see Aaron J. Ihde, "Studies on the History of Rickets. II: The Roles of Cod Liver Oil and Light," *Pharmacy in History* 17, no. 1 (1975), 13–20; also Guy, "History," 115; on "Dreckapotheke," see Aaron J. Ihde, "Studies on the History of Rickets I: Recognition of Rickets as a Deficiency Disease," *Pharmacy in History* 16, no. 3 (1974): 83–88; quote 85.

37. Darbey, quoted in Percival, "Observations," 401; Robert Hunter, "Treatment of Consumption," 11.

38. Guy, "History," 115.

39. Marjorie Cruickshank, *Children and Industry: Child Health and Welfare in Northwest Textile Towns During the Nineteenth Century* (Manchester: Manchester University Press, 1981): 129.

40. The companies were Baker's; Caswell, Massey & Co.; DeClat's Phenate Ammonia; De Jongh's; Ferroleum Emulsion; Fougera's Iodinized; Inland Synthetic; Krohne's Norwegian; Moeller's Norwegian; Morse's; Nichol's; Phillip's Palatable; P.D. & Co's; Quakor; Scott's Emulsion; Stone's; Squibbs; Squire's Emulsion; Wampole's; Waterbury; and Wilbor's; *Meyer Brothers Druggist* 25, no. 12 (December 1904): 52.

41. Within twenty years of its startup in 1873, the New York City company had factories in five countries and marketed Scott's Emulsion on four continents. Diane Wendt, "The Man with a Fish on His Back," Science History Institute, "Distillations," April 2, 2010, https://www.sciencehistory.org/distillations/magazine/the-man-with-a-fish-on-his-back.

42. Advertisement for Scott's Emulsion, *Puck* (December 4, 1889): 228.

43. de Jongh, *Three Kinds*.

44. Ludovicus Josephus De Jongh, *Cod Liver Oil: Causes of its Frequent Inefficacy and Means of Removing the Same; with Remarks upon the Superiority of the Light Brown over the Pale Oil, Directions for its Use and Cases in which the Oil has been used with the Greatest Effect* (London: Ansar, Harford, 1854).

45. This ad claimed Osmun's was "a cure for Consumption, Coughs, Colds, Asthma, Bronchitis, and all Scrofulous Diseases." Ad for Osmun's Cod Liver Oil, *Salt Lake Weekly Tribune*, February 28, 1880, 2; for an Osmun's ad mentioning rickets see *New York Herald*, September 15, 1872, 2, in which "rickets" appears among the "various scrofulous diseases."

46. The ad sold Scott's as "a most valuable food and medicine" for "General Debility, Emaciation, Consumption" or "Wasting in Children." Scotts, "In General Debility, Emaciation, Consumption and Wasting in Children," *Grand Forks Daily Herald*, April 2, 1887, 3.

47. One famous poem is Henry Sambrooke Leigh's "Cod Liver Oil," which begins "On the bleak shore of Norway, I've lately been told, / Large numbers of cod-fish are found, / And the animals' livers are afterwards sold / At so many "pfennigs" per pound; / From which is extracted, with infinite toil, / A villainous fluid called cod-liver oil!" Versions of the "Cod Liver Oil" song are readily available on the internet, including recordings by the Dubliners. Many versions don't get the famous doctor's name right, producing an array of mondegreens, from "doctor, dear John" to "oh doctor, oh dear Dr. John." For a brief introduction to the song, see the page from the University of Maine's Maine Folklife Center: https://umaine.edu/folklife/what-we-do/programs-and-events/maine-song-and-story-sampler-map/places/rumford-cod-liver-oil/ (accessed October 13, 2023).

48. I have paraphrased the bluebird joke, which appears in "Chips," *Hartford Daily Courant*, May 13, 1876, 1; the democracy joke appeared in *Freedom's Champion* (Atchison, Kansas), April 23, 1859, 1; the Delaware man in the *Duluth Minnesotian*, April 4, 1874, 1.

49. Nails and rust, "About People and Things," *Philadelphia Inquirer*, December 3, 1885, 4; "pigeons," *Idaho Statesman* (Boise), June 24, 1876, 3.

50. Notice for "Bliss's Compound Cod Liver Oil Candy. The Original and Only Genuine," *New Hampshire Patriot and State Gazette*, November 21, 1850.

51. Percival, "Observations," Darbey quote at 399, Percival quote at 396; "acidulous liquors" means, essentially, carbonated water.

52. De Jongh, *Three Kinds*, 84.

53. Moller's Cod Liver Oil claimed their mild formula was "more free from unpleasant taste or smell," and could "be taken by persons who cannot retain other oil on their stomach." Ad for "Moller's Cod Liver Oil," *Appleton's Journal of Literature, Science and Art* 7 (March 23, 1872): ii.

54. It may have been a candy, but an 1850 ad promised the candy as an "admirable remedy for coughs, colds, influenza, asthma, bronchitis, and all affections of the throat and lungs." Notice for "Bliss's Compound Cod Liver Oil Candy. The Original and Only Genuine," *New Hampshire Patriot and State Gazette*, November 21, 1850; "Cod Liver Oil and Abolitionism," *Semi-weekly Eagle* (Brattleboro, Vermont), January 23, 1851, 2. According to its byline, the article first appeared in the *Springfield Republican* (the manufacturer B. K. Bliss, Druggist, was based in Springfield, Massachusetts).

55. Rajakumar, Greenspan, Thomas, and Holick, "Solar Ultraviolet Radiation," citing A. F. Hess and M. Weinstock, "Antirachitic Properties Imparted to Inert Fluids and to Green Vegetables by Ultraviolet Radiation," *Journal of Biological Chemistry* 62 (1924): 301–13. Hess wrote to Steenbock that he did "not consider the antirachitic activation of foods by light as being a matter of importance commercially." Quoted from a letter in the University of Wisconsin Archives, in Audrey B. Davis, "The Rise of the Vitamin-Medicinal as Illustrated by Vitamin D," *Pharmacy in History* 24, no. 2 (1982): 59–72, 65.

56. "Tells of New Drug as Rickets Cure," *New York Times*, April 3, 1929, 37; "Discoverer of New Body Building Process Gives It to University to Benefit Humanity," *New York Times*, February 16, 1927, 1.

57. The five companies that received licenses in 1929 were Abbott Laboratories, Mead Johnson, Parke-Davis, Winthrop Chemical Company, and E. R. Squibb. Apple, *Vitamania*, 46.

58. S. S. Oberhelman and T. D. Thatcher, "Vitamin D Deficiency in the 21st Century: An Overview," in Ronald Ross Watson, ed. *Handbook of Vitamin D in Human Health: Prevention, Treatment and Toxicity* (Wageningen, Netherlands: Wageningen Academic Publishers, 2013), 16.

59. Apple, *Vitamania*, 38–40.

60. A list of WARF licensees appears in "Vitamin D Patents Given to the Public," *New York Times*, January 15, 1946, 16.

61. Apple, "Patenting University Research," 387.

62. "Keep Sunny Summer Health—Drink Schlitz All Winter," *Esquire*, November 1936, 175.

63. Fleischman's yeast received a license from WARF in 1928,

74th Congress, House Committee on Patents, *Pooling of Patents. Hearings. . . . Feb. 11-Mar. 7, 1935*, 3549.

64. Apple, *Vitamania*, 50; although Fleishman's had a license from WARF in 1928, the ad refers to "Schlitz Yeast" as if it is a brand name.

65. Ruth Dupré, "'If It's Yellow, It Must Be Butter': Margarine Regulation in North America Since 1886," *Journal of Economic History* 59, no. 2 (1999): 353–71; James Harvey Young, *Pure Food: Securing the Federal Food and Drugs Act of 1906* (Princeton, NJ: Princeton University Press, 1989), esp. chapter 4, "This Greasy Counterfeit," 66–94.

66. F. B. Morrison to Harry L. Russell, December 14, 1925, WARF files, quoted in Apple, "Patenting University Research," 378.

67. One sign of oleomargarine's changing stature was its manufacturers' ability to unite. Within seven years of its founding in 1920, the Institute of Margarine Manufacturers represented the makers of 85 percent of America's margarine. Richard A. Ball and J. Robert Lilly, "The Menace of Margarine: The Rise and Fall of a Social Problem," *Social Problems* 29, no. 5 (June 1982): 494.

68. Great Britain chose an entirely different, and margarine-friendly, path to vitamin D fortification. In the 1930s, the fear that rickets might rebound among poor British children was a regular topic in Parliament and was central in planning food supply issues in the coming war. The government mandated vitamin D fortification of all margarine and "the National Dried Milk" in the 1940s "based on the dietary habits of the majority population, in combination with the rationing of butter," and in accordance with long-standing pure milk laws. Bivins, "'English Disease,'" 545n38.

69. Meghan Crnic and Michelle C. Kondo, "Nature Rx: Reemergence of Pediatric Nature-Based Therapeutic Programs from the Late 19th and Early 20th Centuries," *American Journal of Public Health* 109 (2019):1371–78, https://doi.org/10.2105/AJPH.2019.305204). On city campaigns to increase light in buildings and streetscapes, see Freund, *American Sunshine*, esp. 7–36; the literature on parks and playgrounds in the Gilded Age and Progressive Era is extensive. A fine place to start is Dom Cavallo, *Muscles and Morals: Organized Playgrounds and Urban Reform, 1880–1920* (Philadelphia: University of Pennsylvania Press, 1981). The quote is from Woods Hutchinson, "A Time to Be Out of Doors," *North American Review* 155, no. 429 (August 1892), 253–56, quote 254.

70. David Stradling, *Smokestacks and Progressives: Environmentalists, Engineers, and Air Quality in America, 1881–1951*, revised edition (Baltimore: Johns Hopkins University Press, 2002); Frank Uekötter, *The Age of Smoke: Environmental Policy in Germany and the United States, 1880–1970* (Pittsburgh: University of Pittsburgh Press, 2009).

71. Edgar Mayer, "Sun-Bathing and Sun-Lamps," *American Journal of Nursing* 33, no. 8 (August 1933), 739–46, quote 747.

72. "How fortunate," "peace of mind": ad for "Hanovia Alpine Sun Lamp," *Hygeia*, March 1929, 211; "Robust, suntanned," *Hygeia*, May 1928, 5.

73. "Bathed in Living Light!," Ad for "Vita Glass," [No publication title given, but shown as January 1929], 97, Grace's Guide to British Industrial History, accessed October 20, 2023, https://www.gracesguide.co.uk /File:Im192901GHK-VitaGlass.jpg.

74. Freund, *American Sunshine*, 76; John Stanislav Sadar, *Through the Healing Glass: Shaping the Modern Body through Glass Architecture, 1925–35* (New York: Routledge, 2016): 220–22.

75. Cancer protection plays a smaller role in modern window-glass marketing than one might expect. One manufacturer, American Window Products, initially invites customers to "think about your skin if you sit outside under the sun . . . with no protection." They then assure that their "new windows'" UV filtering glass offers the needed protection of "home assets," preventing damage "to your furniture, flooring, as well as other items." "The UV Protection New Windows Provide and How It Protects Home Assets," American Window Products, accessed, August 9, 2022, https://www .americanwindowproducts.com/the-uv-protection-new-windows-provide -and-how-it-protects-home-assets/; a similar website for PPG's "Vitro Architectural Glass" makes no mention of health benefits. "SUNGATE 500 Passive Low-E Glass," Vitro Architectural Glass, accessed August 9, 2022, http://www.vitrowindowglass.com/lowe_glass/sungate_500.aspx.

Chapter 6

1. "Milk Irradiation for Rickets Urged," *New York Times*, November 11, 1933, 17.

2. Edwards A. Park to Martha May Eliot, July 15, 1934, Edwards A. Park Collection, Folder "Eliot, Martha 1933–34," Chesney Archives, Johns Hopkins Medicine, Nursing, and Public Health.

3. Smith-Howard, *Pure and Modern Milk*, 12–35.

4. Melanie DuPuis, *Nature's Perfect Food: How Milk Became America's Drink* (New York: New York University Press, 2002).

5. G. C. Supplee, "Antirachitic Activation of Milk by Direct Irradiation with Ultra-Violet Rays," *AJPH* 23, no. 3 (March 1, 1933): 225–29; quote 226.

6. DuPuis, *Nature's Perfect Food*; Marc T. Law, "The Origins of State Pure Food Regulation," *Journal of Economic History* 63, no. 4 (December 2003): 1103–30.

7. Harry L. Russell, director of WARF, to L. D. Johnson, of Mead Johnson, March 21, 1934, folder "Mead Johnson and Company, 1933–1939," Edwards A. Park Collection, Chesney Archives, Johns Hopkins Medicine, Nursing, and Public Health. In a letter to Edwards A Park, W. N. Larson, VP at Mead Johnson, expressed their intention to forego milk fortification, focusing instead on vitamin D concentrates that parents would administer directly to their children. Larson to Edwards A. Park, 31 May, 1934, folder "Mead Johnson and Company, 1933–1939," Edwards A. Park Collection, Chesney Archives, Johns Hopkins Medicine, Nursing, and Public Health.

8. Benjamin Schwab, "Nutrition, Education and Development: The Case of Vitamin D Milk," paper prepared for presentation at the Agricultural & Applied Economics Association's 2011 AAEA & NAREA Joint Annual Meeting, Pittsburgh, Pennsylvania, July 24–26, 2011, 9, https://ageconsearch.umn .edu/record/103614/?ln=en.

9. John W. Gowen, J. M. Murray, M. E. Gooch, and Forrest B. Ames, "Rickets, Ultra-Violet Light and Milk," *Science* 63, No. 1621 (January 22, 1926), 97–98. A German experiment conducted the same year found increased and potentially curative levels of vitamin D in the milk of irradiated cows. C. Falkenheim, W. Voltz, and W. Kirsch, "Quartzlampen Bestrahlung von Kühen und antirachitische Wirksamkeit ihrer Milch," *Klinische Wochenschrift* 5 (1926): 2071.

10. Mitchell, Eiman, Whipple, and Stokes, "Protective Value."

11. Alfred S. Hess, Mildred Weinstock, and Elizabeth Sherman, "Production of Antirachitic Properties in Human Milk Resulting from Irradiation of Mother," *Proceedings of the Society for Experimental Biology and Medicine* 23 (1926): 636.

12. James A. Tobey, "Vitamin D Milks," *American Journal of Nursing* 36, no. 6 (June 1936): 576–80, 577.

13. In 1935, 212 US dairies were producing metabolized vitamin D milk from 5,002 cows. "Metabolized Vitamin D Milk Now Widely Available," *Vitamin D Milk* 2, no. 2 (February 1935): 11; Schwab, "Nutrition, Education," 9.

14. Wisconsin Alumni Research Foundation, "Get the Sunshine In: The Licensing of Irradiated Milk in 1932," accessed November 3, 2023, https:// www.warf.org/announcement/get-the-sunshine-in-the-licensing-of -irradiated-milk-in-1932/.

15. "Over 8 years" and "a method similar to," Borden's ad, *New York Times*, October 5, 1935, 6; "Yes folks," Borden's ad, *New York Times*, February 17, 1939, 8.

16. "A Standing Invitation," Ad for Willow Brook Irradiated Vitamin D Milk, *Scarsdale Inquirer*, January 4, 1935, 7. According to the information

cards accompanying the exhibit, the irradiator was made by the Creamery Package Manufacturing Company, "A Milk Irradiator Used to Put Vitamin D in Milk," image COP_17_0001_00011_007, Century of Progress records, Special Collections and University Archives, University of Illinois at Chicago Library, accessed November 3, 2023, https://collections.carli.illinois.edu /digital/collection/uic_cop/id/365.

17. The dairies turned the question of quality control to their advantage: a Borden ad celebrated that "A Westinghouse Recording Ultra-Violet Ray Meter controls and records the Vitamin D potency during the process of irradiation." *Scarsdale Inquirer*, September 13, 1935, 3.

18. Apple, "Patenting University Research," 387–88; Kenneth G. Weckel, "Vitamin D in Milk—A Review" *Journal of Dairy Science* 24, no. 5 (May 1941): 445–62, 455.

19. Matthias Wacker and Michael F Holick, "Sunlight and Vitamin D: A Global Perspective for Health," *Dermatoendocrinology* 5, no.1 (2013): 51–108, https://doi.org/10.4161/derm.24494. Yet an industry newsletter in 1935 admitted that only around 3 percent of liquid milk sold in the States was fortified, "Amount of Vitamin D Milk Now Used Constitutes Important Percentage of Total," *Vitamin D Milk*, October 1936, 28.

20. Tobey, "Vitamin D Milks," 577.

21. Market penetration numbers are scarce. My calculation, based on contemporary reports in the *AJPH* and historic consumption data from the USDA estimated penetration at only 2 percent; quote from Kenneth G. Weckel, "Control Practices Used in Supervision of Vitamin D Milk by City and State Milk Sanitarians," *Journal of Milk and Food Technology* 10, no. 3 (1947): 167–71, quote 167.

22. Dr. Chauncey D. Leak, professor of pharmacology at the University of California, San Francisco, quoted in "Warn of Stressing Vitamin D in Foods," *New York Times*, November 23, 1936, 16.

23. American Public Health Association Committee of the Food and Nutrition Section, "Milk and Dairy Products," APHA Eighth Annual Year Book, AJPH 28 no. 2 (suppl. 56): 66–75, esp. 67.

24. "Vitamin D Milk Approved," *New York Times*, March 15, 1935, 44.

25. Starting in 1935, "every week-day night (except Saturday)" New York's Sheffield Farms sponsored a fifteen minute children's adventure radio program, *Billy and Betty*, starring Jimmy McCallion and Audrey Egan. Terry G. G. Salomonson, "Information and Help to the New Collector - Part VI/ Serials: (Part 1) – Juvenile Radio Programs," accessed November 3, 2023, http://www.otrsite.com/articles/art1006.html; see also their ad in the *Rockland County Journal News*, June 19, 1935. Dairies (and WARF) also marketed directly to doctors and sponsored public lectures, often

by WARF-affiliated scientists. The *New York Times* reported on a WARF lecturer, Henry T. Scott of Madison Wisconsin, speaking at a luncheon of the Women's City Club, "Banishing of Rickets by Milk Diet Seen," *New York Times*, March 24, 1935, N2.

26. Sheffield ad, *New York Times*, May 29, 1935, 22; Borden's ad, *New York Times*, May 29, 1935, 8; that *"taking"* might seem to convey an especially medicinal tone, but it is more likely just a convention from the days of doorstep delivery. Willow Brook ad, *Scarsdale Inquirer*, January 1, 1935, 7. It appears that Westchester County dairies were test-marketing irradiated milk six months before dairies in New York City.

27. On the price assurance to professionals, Carnation Company drafted a form letter to be sent to physicians, 8 June 1934, folder "Milk, Wisconsin Fund," folder #3, Joseph Stokes Papers, American Philosophical Society, Philadelphia B: St65P. Sheffield ad, *New York Times*, May 29, 1935, 22; Borden's ad, *New York Times*, May 29, 1935, 8.

28. Borden's ad, *Scarsdale Inquirer*, September 13, 1935, 3; "you cannot," Willow Brook ad *Scarsdale Inquirer*, January 4, 1935, 7; Sheffield Farms promised their milk "helps take the place of summer sun all winter," Sheffield ad, *New York Times*, September 12, 1935, 12; Sheffield ad, *New York Times*, November 7, 1935, 18.

29. "You Don't Want," *New York Amsterdam News*, March 25, 1939, 7; "LEGS tell the story," Sealtest ad, *New York Times*, January 15, 1937, 44. The ads went beyond rickets prevention, targeting the "whole family," especially women. A 1937 Borden ad highlighted the dental benefits of vitamin D milk by recalling an old saying that "a woman lost a tooth for every child." The benefits to women extended beyond childbearing years. A Sheffield ad for Mother's Day, 1941, explained "why mothers are no longer old at 40." The ad recommended women over forty should drink a quart each day. Borden's ad, *New York Times*, March 12, 1937, 48. The ad also directed women to tune their radios to "Special Edition," which an industry newsletter described as news commentary "aimed at the femme listeners," with "information for housewives and hostesses." *Radio Daily*, September 21, 1937, 6, https://worldradiohistory.com/hd2/IDX-Business/Magazines/Archive -Radio-Daily-IDX/RD-37/RA-1937-sep-OCR-Page-0110.pdf. "Mothers are no longer old at 40," *New York Times*, May 7, 1941, 22.

30. "D-Lighted to Meet You," *New York Times*, January 17, 1939, 16. The cow that would become Elsie first appeared in a series of medical journal ads in 1936. Christine Bunish, "Borden, Inc." in John McDonough and Karen Egolf, *The Advertising Age Encyclopedia of Advertising* (New York: Routledge, 2015): 192–96. Borden's ads continued to feature anthropomorphic bovines in a range of situations: impersonating Mae West,

"Come Down and See Me Sometime," *New York Times*, January 3, 1939, 10 (here the cow impersonating Mae West and flirting with the sun is named Nellie); or posing as an old lady cow to link Social Security to the care of Borden's herds, "Some Cows Have Social Security," *New York Times*, February 20, 1939; 17. One ad featured the Dionne quintuplets: "I've been adopted by the quints!" the cow boasted. Their doctor wanted "those five famous girls up in Canada" to get plenty of vitamin D, and Borden was providing it in "the most pleasant way." "I've Been Adopted by the Quints," *New York Times*, May 23, 1939, 15.

31. "Elsies's in the Mo-od for Love," *New York Times*, November 29, 1939, 13. Over the next years, Elsie's bovine face was rendered more realistically, but the rest of her acquired more human features until she appeared as an American housewife with a cow's head, crossing into some cartoon animal version of the human/android discomfort factor known as the "uncanny valley."

32. "Cod Liver Oil Supply Reported Adequate," *New York Times*, September 25, 1940, 54; USDA Economic Research Service, "Table 124.— Fish Liver Oil (Including Cod Oil): Supply and Disposition, 1912–65)," "US Fats and Oils," Statistical Bulletin No. 376, 1966, 109.

33. For an in-depth analysis of vitamin advertising in the 1920s and 30s, see Apple, *Vitamania*, 13–32; for an overview of cod liver oil advertisements of the 20s and 30s, see Freund, *American Sunshine*, 145–50; "In Winter Months They Get So Little Sunshine," *Hygeia*, October 1928, 22; Rima Apple cites the same ad, but in a different magazine, Ad for Squibb's cod-liver oil, *McCall's*, November 1928, 61 (reprinted in Apple, *Vitamania*, 269).

34. *Good Housekeeping*, December 1, 1934, 223.

35. Taken from Coco Cod label, Indiana State Archives Facebook page, accessed November 3, 2023, https://www.facebook.com/INStateArchives /photos/1673429609491329; another ad assured Coco Cod was "the only kind that actually tastes good and does not reek of fish!" *Chicago Tribune*, October 27, 1929, 9. In 1911 Eli Lilly had sold their own chocolate-flavored cod liver oil. "Coco-Emulsion of Cod Liver Oil," *N.A.R.D. Notes: Journal of the National Association of Retail Druggists* 11 (1911): 315; ad for Nyal Wild Cherry and Cod Liver Compound, *Ramapo Valley Independent*, October 21, 1927, 18.

36. White's claimed to have removed the "nauseating fishy fats," producing a "modernized pleasant way" to administer cod liver oil. "Cod Liver Oil Nightmare," *Good Housekeeping*, December 1, 1933, 160.

37. For Squibb, see Freund, *American Sunshine*, 147; "A Picture of Health," *Good Housekeeping*, December 1, 1934, 180.

38. "Convalescing?" ad for Scotts Emulsion, *New York Amsterdam News*, September 16, 1944, 8A; Scott's marketed heavily in African American media. The company's ads in *Ebony* magazine into the late 1950s usually featured black children and highlighted Scott's power to help recover from an existing cold and prevent new ones. They rarely if ever mention rickets. See, for example, "One Cold After Another?," *Ebony*, 1957 (no month given), Duke University Libraries "Medicine and Madison Avenue," accessed November 3, 2023, https://repository.duke.edu/dc/mma/MM0355.

39. "Since Viking Days," ad for Scott's Emulsion," *New York Times*, December 10, 1922, 34; Squibb ad, *New York Times*, March 22, 1925, 124.

40. "Bottled Sunshine," *New York Times*, October 2, 1927, RP5. Another ad in the "Bottled Sunshine" campaign depicted a bottle of Squibb's brilliantly backlit, radiating beams of sunlight in all directions, similar to the depictions of UV sunlamps, whose popularity was peaking at about the time, *Hygieia* October 1928: 22.

41. "Health" by Royal S. Copeland, MD, *The Oregon Statesman*, June 21, 1934, 4; Children's Bureau quoted in Golden, *Babies*, 114.

42. Golden, *Babies*, 114–15.

43. "for infants who refuse," H. C. Cameron, "Fifty Years of Progress in Paediatrics," *The Clinical Journal* 70 no. 12 (December 1941): 309–15, quote 315; "would still prescribe," The "extensive survey of American pediatricians" is described in a 1937 report from American Public Health Association's section on Food and Nutrition. APHA, "Milk and Dairy Products," 68.

44. "tell your patients," "Stone Walls Do ~~Not~~ a Prison Make," Mead's ad, *The Nova Scotia Medical Bulletin* 11, no. 3 (March 1932): 156; David H. Shelling and Katherine Hopper, "Calcium and Phosphorus Studies. XII. Six Years' Clinical Experience with Viosterol in the Prevention and Treatment of Rickets, Tetany, and Allied Diseases," *Bulletin of the Johns Hopkins Hospital* 63, no. 3 (March 1936): 153.

45. A typical example is a full-page color 1948 ad for White's Cod Liver Oil Concentrate Liquid, which makes a headline of the quote "by the tenth or twelfth day of life . . . vitamin D in a concentrated form should be started," with a footnote calling to an article in the *Pennsylvania Medical Journal*, *Journal of the Florida Medical Association* 34 (1947–8): insert between 174–175; "Both Vitamins Definitely Measured," ad for Parke Davis Standardized Cod-Liver Oil, *Journal of the Florida Medical Association* 16 (1929–30): 325.

46. Mead was one of five pharmaceutical companies to which WARF granted licenses to manufacture Viosterol in 1929. "Summer-Time Use of

Viosterol," Mead Johnson & Co. ad, *Journal of the Medical Association of the State of Alabama* 1 no. 1 (July 1931): 1.

47. "Ethically Marketed," "How Much Sun Does the Baby Really Get?," ad for Mead's Oleum Percomorphum, *Bulletin of the Los Angeles County Medical Association* 68 (November 3, 1938): 798; "face to face," White's ad, *American Druggist* 101 (January 1940): 107. On the broader history of ethical marketing, see Julie Donohue, "A History of Drug Advertising: The Evolving Roles of Consumers and Consumer Protection," *The Milbank Quarterly* 84, no. 4 (December 2006): 659–99. On AMA's self-regulation playing out with rickets prevention, see "Prescribing of Antirachitics," *Southern Medical Journal* 23, no. 4 (April 1930): 366–67; "Deeds, Not Words," Mead ad on ethical marketing, *Rhode Island Medical Journal* 15, no. 268 (April 1932): II.

48. NIH Office of Dietary Supplements, "Vitamin D—Health Professional Fact Sheet," quoted in Julie R. Thomson, "The Real Reason the Dairy Industry Adds Vitamin D to Your Milk," *Huffington Post*, March 29, 2017, https://www.huffpost.com/entry/why-vitamin-d-in-milk_n_58da 7a68e4b0286e65b5de62; recent revisions of this fact sheet are far more nuanced, even giving partial credit to cod liver oil: "The fortification of milk (a good source of calcium) and other staples, such as breakfast cereals and margarine, with vitamin D beginning in the 1930s along with the use of cod liver oil made rickets rare in the United States." NIH Office of Dietary Supplements, "Vitamin D—Fact Sheet for Health Professionals," accessed November 3, 2023, https://ods.od.nih.gov/factsheets/VitaminD-Health Professional/; Rajakumar, "Vitamin D, Cod-Liver Oil."

49. Weick, "History of Rickets." Her essay, covering rickets from colonial days to the mid-twentieth century, was one of the first such surveys, and has been cited in many dozen books and journal articles. The article is her only publication cited in *Medline*.

50. The US mortality data for 1910 to 1961 appear in Weick, "History of Rickets." Subsequent reports show that the long tail extended from 1962 to 1981, with under two deaths per year on average. US Department of Health, Education and Welfare, Vital Statistics of the United States (US: GPO, 1962–1981), https://www.cdc.gov/nchs/.

51. Fred O. Tonney, "Vitamin D in Child Health" *AJPH* 26, no. 7 (July 1936): 665–71, quote 669.

52. Julius Levy and Harry B. Silver, "Can Rickets Be Eliminated from the Large Cities?" *Archives of Pediatrics* 56 (1939): 96–105, quote 104, 105.

53. Edwards A. Park, *Dental Cosmos*, February 1923, quoted in "The Prevention of Rickets," *AJPH* 16, no. 2 (February 1926): 139–41.

54. Weckel, "Control Practices," 167.

55. Eileen B. Yeh, David M. Barbano, and MaryAnne Drake, "Vitamin Fortification of Fluid Milk," *Journal of Food Science* 82, no. 4 (2017): 856–64.

56. On WARF developing UVO see Apple, "Patenting University Research," 387–88; in the early 1940s WARF still promoted the three WARF-sanctioned methods of fortifying milk: irradiation, the addition of UVO, and "metabolized" milk from cows fed irradiated yeast. "School Days Bring Increased Need for Vitamin 'D' Milk," *American Journal of Public Health* 31 (October 1941): ad section, V.

57. B. R. East, "Another Job for the Codfish," *Scientific American* 150 (May 1934): 244–45, quote 245.

58. Julie A. Burger and Justin Brunner, "A Court's Dilemma: When Patents Conflict with Public Health," *Virginia Journal of Law & Technology* 12, no. 7 (Fall 2007): 1–42; "Vitamin D Patents Given," 16.

59. Yeh, Barbano, and Drake, "Vitamin Fortification"; Anthony W. Norman, "Vitamin D and Milk," accessed February 14, 2015, http://vitamind.ucr.edu/about/ [dead link in 2021, but captured on the Wayback Machine, https://web.archive.org/web/20120608061309/http://vitamind.ucr.edu/about/, accessed October 20, 2021]; "Vegan" vitamin D3 is now available using a process of fortifying lichen.

60. On discovery of vitamin D3, see Hector F. DeLuca, "History of the Discovery of Vitamin D and Its Active Metabolites," *BoneKEy Reports* 3, article number 479 (2014): 1–9, https://doi.org/10.1038/bonekey.2013.213; D3's greater efficacy was first proved in animal experimentation but was established in human physiology as well. C.A. Elvehjem, "Recent Advances in Our Knowledge of the Vitamins," *Scientific Monthly* 56, no. 2 (February 1943): 99–104; K. Glaser, A. H. Parmalee, and W. S. Hoffman, "Comparative Efficacy of Vitamin D Preparations in Prophylactic Treatement of Premature Infants," *American Journal of Diseases of Children* 77, no. 1 (1949):1–14, https://doi.org/10.1001/archpedi.1949.02030040008001.

61. Markets and Markets, "Vitamin D Market," accessed November 3, 2023, https://www.marketsandmarkets.com/Market-Reports/vitamin-d-market-22034298.html.

62. Vitamin D supplement sales rose from $42 million in 2002 to $936 million in 2017. For 2002 to 2011, see Katharine Greider, "Has Vitamin D Been Oversold?" *AARP Bulletin*, July 5, 2012, http://www.aarp.org/health/drugs-supplements/info-07-2012/how-much-vitamin-d-is-enough.html; for 2017 figure, see Liz Szabo, "Selling America on Vitamin D—And Reaping the Profits," *NBC Health News*, August 20, 2018, https://www.nbcnews.com/health/health-news/selling-america-vitamin-d-reaping-profits-n902276.

63. Peak imports were 67 million pounds in 1937. In 1999, the US imported only 2.6. In truth, imports were increasing in 1999; that 2.6 million pounds was up from 2.2 million just five years earlier. Source: USDA Economic Research Service, "US Fats and Oils," 109; National Oceanic and Atmospheric Administration, *Imports and Exports of Fishery Products Annual Summary*, 2000, 8, https://www.st.nmfs.noaa.gov/st1/trade/documents/trade2000.pdf.

64. In the early twenty-first century, cod liver oil imports more than doubled to a new plateau of around $10 million. US Census Bureau: Economic Indicators Division, "Cod Liver 92 to 2021," USA Trade Online, " https://usatrade.census.gov/.

65. "Reviews and Bibliographic Notices" [a review of six books on cod liver oil], *Dublin Journal of Medical Science* 9, no. 18 (1850): 412–52, quote 418.

Chapter 7

1. The puff piece did offer a qualification, mentioned below. Anna May Wilson, "Food and Health: The Conquest of Rickets: The Disease of Poverty and Darkness," *Today's Health* 30, no. 5 (May 1952): 44–45. *Today's Health*, published by the American Medical Association, was the successor publication to the AMA's long-running *Hygeia*.

2. Through the 1950s and 1960s, Harold Harrison and Helen Harrison coauthored dozens of cutting-edge papers on vitamin D metabolism; H. E. Harrison, "The Disappearance of Rickets, *AJPH* 56, no. 5 (May 1966): 734–7; quotes 734.

3. Wilson, "Food and Health: The Conquest of Rickets," 45; quote from "Unraveling the Enigma of Vitamin D," in *Beyond Discovery: The Path from Research to Human Benefit* (Washington: National Academy of Sciences, 2000), http://www.nasonline.org/publications/beyond-discovery/vitamin-d.pdf.

4. DeLuca, "History," 2; Nicole Kresge, Robert D. Simoni, and Robert L. Hill, "A Half-Century of Vitamin D: the Work of Hector F. DeLuca," *Journal of Biological Chemistry* 281, no. 50 (December 15, 2006): e41–43.

5. Harold E. Harrison and Helen C. Harrison, "Rickets Then and Now," *Journal of Pediatrics* 87, no 6. (December 1975): 1144–51, quote 1144.

6. For a blow-by-blow report of the research, with citations to all the important papers, see DeLuca, "History," 2.

7. On the various forms of rickets, see L. S. Nield, P. Mahajan, A. Joshi, and D. Kamat, "Rickets: Not a Disease of the Past," *American Family*

Physician 74, no. 4 (2006): 619–26; Harrison and Harrison, "Rickets Then and Now."

8. David Krestin, "The Prophylaxis of Rickets by Single Massive Doses of Vitamin D," *British Medical Journal* 1, no. 4385 (January 20, 1945): 78–80.

9. Germany made sure to let the western media know about their efforts. For example, "Reich to Combat Rickets," *New York Times*, January 19, 1940, 6; "Reich Beating Rickets," *New York Times*, February 4, 1940, 22. In 1941, the Reich Ministry of Health released a slick thirteen-minute film, *Die englishe Krankheit*. The film concludes with a mother feeding her child cereal fortified with Vigantol, a vitamin D supplement from Bayer and Merck. The narration promises that with a few drops of Vigantol every day, National Socialist Germany would be sure that the English Disease would never become a German disease. For a detailed overview of the film, see Anja Laukötter, " 'Light, Air, and Sun!' Die englische Krankheit [The English Disease] and Health Education Films in the Third Reich," an essay in the National Library of Medicine's "Medicine on Screen" series, accessed June 3, 2021, https://medicineonscreen.nlm.nih.gov/2015/12/22/die-englische -krankheit-the-english-disease/.

10. The first alarm concerning cases of hypercalcemia appeared in 1953, when clinicians reported ten infants with hypercalcemia showing similar symptoms, including anorexia, vomiting, wasting, and constipation. Harry Bakwin, "The Overuse of Vitamins in Children," *American Journal of Pediatrics* 59, no. 1 (July 1962): 154–61; "Infantile Hypercalcaemia, Nutritional Rickets, and Infantile Scurvy in Great Britain," *British Medical Journal* 1, no. 5399 (June 27, 1964): 1659–61; "Vitamin D as a Public Health Problem," *British Medical Journal* 1, no. 5399 (June 27, 1964): 1654–55; for the postwar hypercalcemia concerns in context, see Bivins, "Ideology and Disease."

11. "The Cause of the Children, *New York Times*, August 13, 1947, 22. *Seeds of Destiny*, a twenty-minute documentary, was produced for the wartime relief agency, the United Nations Relief and Rehabilitation Administration. U.S. Army Signal Corps, *Seeds of Destiny* (US Department of War, 1946).

12. Weick, "History of Rickets."

13. The survey she cited is described in detail in Fred Brooksaler, "Rickets Due to Vitamin D Deficiency," *Clinical Pediatrics* 1, no. 1 (October 1962): 57–61.

14. Harrison, "Disappearance," quoted in Shelley R. Kreietr et al, "Nutritional Rickets in African American Breast-Fed Infants," *Journal of Pediatrics* 137, no. 2 (August 2000): 153–57, https://doi.org/10.1067/mpd .2000.109009; Weick, "History of Rickets."

15. Douglas Crawford McMurtrie, "The Care of Crippled Children in the United States," *American Journal of Orthopedic Surgery* s2–9, no. 4 (May 1912): 527–56, 527.

16. Henry T. Ireys, Robert J-P. Hauck, and James M. Perrin, "Variability among State Crippled Children's Service Programs: Pluralism Thrives," *American Journal of Public Health* 75, no. 4 (April 1985): 375–81; Jeffrey Paul Brosco, "Navigating the Future Through the Past: The Enduring Historical Legacy of Federal Children's Health Programs in the United States," *American Journal of Public Health* 102, no. 10 (October 1, 2012): 1848–57; Janet Golden and Jeffrey P. Brosco, "The United States Children's Bureau and Pediatric Medicine: A Retrospective Analysis," *Pediatrics* 130, no. 6 (December 2012): 993–95, https://doi.org/10.1542/peds.2012-1865.

17. Sister Michael Joyce, "Assignment: East Harlem," *American Journal of Nursing* 69, no. 7 (July 1969), 1497–1502.

18. Mrs. G. had developed eclampsia during the premature delivery of her last child, named "Tiny." "Last child" because "at the time of delivery a tubal ligation had been done." Tiny, now eleven pounds, was "malnourished and underdeveloped" and had just come home from the hospital where he had been treated for dehydration. Joyce, "Assignment: East Harlem," 1497.

19. Joyce, "Assignment: East Harlem." "perfect 'O,'" 1498; "grow out of his crooked legs," 1500.

20. Joyce, "Assignment: East Harlem," 1502.

21. William Farnsworth Loomis, "Rickets," *Scientific American* 223, No. 6 (December 1970): 76–91.

22. For example, a comparative study of data from a Public Health Service survey of several communities—some predominately Black, some white, and one county with sizable Black and white populations—showed rates of malnutrition and rickets in remote rural white counties in Michigan that rivaled those found in predominately Black inner-city communities near Detroit. "The Health Status and Health Education of Negroes in the United States," *Journal of Negro Education* 18, no. 3 (Summer 1949), 291–304.

23. Joyce, "Assignment: East Harlem," 1500.

24. "pot-liquor," "Have We any Rickets?" Note that rickets could affect pelvic growth in several ways; obstetricians have catalogued over a half-dozen unique pelvic distortions that can impede childbirth. The label "rachitic pelvis" named a particular "flat" pelvis closely associated with childhood rickets, but rickets is usually listed as a potential cause for several of the others. William T. Lusk, *Science and Art of Midwifery* (New York: D. Appleton, 1884): 433; Harrison's point was that rickets caused not only the classic symptoms ("hypocalcemic tetany" and visible deformities) "but

also the hidden deformity of the pelvis which in the female was a major cause of dystocia and almost certainly of maternal and infant mortality and morbidity." Harold E. Harrison, "Rickets," *Western Journal of Medicine* 125 (September 1976): 223–25; quote 223, PMCID: PMC1237284.

25. Peter M. Dunn, "The Chamberlen Family (1560–1728) and Obstetric Forceps," *Archives of Disease in Childhood Fetal and Neonatal Edition* 81 (1999): F232–F235; see Randi Hutter Epstein, *Get Me Out: A History of Childbirth from the Garden of Eden to the Sperm Bank* (New York: Norton, 2010) for a spirited telling of the Chamberlens' story.

26. Alban Doran, "A Chronology of the Founders of the Forceps," *Journal of Obstetrics and Gynaecology* 27 (1915): 154–72; Wendy Moore, "Keeping Mum," *British Medical Journal* 334, no. 7595 (March 31, 2007): 698.

27. Adrian Wilson, *The Making of Man-Midwifery: Childbirth in England, 1660–1770* (Cambridge, MA: Harvard University Press, 1995), 57.

28. The evolutionary process described here is one aspect of the theory of the "obstetrical dilemma," a term coined by physical anthropologist Sherwood Washburn in 1960 to describe a hypothesis nearly as old as Darwinism. Recently critics have chipped away at some of the details of Washburn's hypothesis, but new theories have arisen to bolster Washburn's general thesis. M. Haeusler, N. D. S. Grunstra, R. D. Martin, V. A. Krenn, C. Fornai, and N. M. Webb, "The Obstetrical Dilemma Hypothesis: There's Life in the Old Dog Yet," *Biological Reviews of the Cambridge Philosophical Society*, May 19, 2021, Epub ahead of print, https://doi.org/10.1111/brv .12744, PMID: 34013651.

29. Peter M. Dunn, "Dr Percivall Willughby, MD (1596–1685): Pioneer 'Man' Midwife of Derby," *Archives of Disease in Childhood* 76 (1997): F212–13.

30. He married a midwife with whom he had ten children. David P. Steensma and Robert Kyle, "Hendrik van Deventer: Dutch Obstetrician and Orthopedist," *Mayo Clinic Proceedings* 91, no. 12 (December 1, 2016): e171.

31. Hendrik van Deventer, *Manuale operatien, Ie deel, zijnde een nieuw ligt voor vroedmeesters en vroed-vrouwen, haar getrouwelijk ontdekkende al wat nodig is te doen, om barende vrouwen te helpen verlossen* (Gravenhage: by the author, 1701), 5; Hendrik van Deventer, *Operationes chirurgicae novum lumen exhibentes obstetricantibus, quo fideliter manifestatur ars obstetri- candi et quidquid ad eam requiritur* (Lugduni Batavorum (Leiden): Andreas Dyckhuisen, 1701).

32. That honorific often goes to William Harvey, the seventeenth-century physician and physiologist most famous for his study of the workings of the heart and blood vessels. See, for example, R. W. Johnstone,

"William Harvey—'The Father of British Midwifery,'" *British Journal of Obstetrics and Gynecology* 55 (1948): 293–302.

33. The full title of the atlas is *A Sett of Anatomical Tables, with Explanations, and an Abridgement of the Practice of Midwifery, with a View to Illustrate a Treatise on the Subject and Collection of Cases*, accessed October 20, 2023, https://anatomia.library.utoronto.ca/islandora/object/anatomia%3 ARBAI042.

34. WHO Newsroom, "Obstetric Fistula," accessed November 3, 2023, https://www.who.int/news-room/facts-in-pictures/detail/10-facts-on -obstetric-fistula.

35. "anything I hated," J. Marion Sims and H. Marion Sims, eds., *The Story of My Life* (New York: D. Appleton: 1884), 231; A local physician, Dr. Henry, called Sims in to assist with Anarcha's difficult birth, suggesting he bring his instruments. She had been in labor for three days without progress. The "child's head was so impacted in the pelvis that the labor-pains had almost entirely ceased." He and Dr. Henry agreed "the sooner she was delivered the better, and without any great effort the child was brought away with forceps," a turn of phrase that suggests—though Sims never confirmed—that Anarcha's child had not survived (226–227).

36. Sims and Sims, *Story of My Life*, 236; see also Dierdre Cooper Owens, *Medical Bondage: Race, Gender, and the Origins of American Gynecology* (Athens: University of Georgia Press, 2017), 34–38.

37. L. Lewis Wall, "Did J. Marion Sims Deliberately Addict His First Fistula Patients to Opium?," *Journal of the History of Medicine and Allied Sciences* 62, no. 3 (2007): 336–56, https://doi.org/10.1093/jhmas/jrl045.

38. George Gray Ward, "Marion Sims and the Origin of Modern Gynecology," *Bulletin of the NYAM* 12, no. 3 (March 1936): 93–104; Deborah Kuhn McGregor, *From Midwives to Medicine: The Birth of American Gynecology* (Camden, NJ: Rutgers University Press, 1988), 109; Owens, *Medical Bondage*, 96.

39. The bodies were reinterred on Ward's Island. Ward, "Marion Sims," 98.

40. "The Woman's Hospital, Running Notes of its History, Progress and Present Efficiency," *New York Herald*, June 19, 1868, 8. See also McGregor, *Midwives to Medicine*, 79–80.

41. McGregor, *Midwives to Medicine*, 110.

42. McGregor, *Midwives to Medicine*, 110; Lusk's quote appears in Lusk, *Science and Art*, 433.

43. For a reconsideration of women's roles in these developments, see Carrie Pauline Adkins, "'The Sacred Domain': Women and the Transformation of Gynecology and Obstetrics in the United States, 1870–1920" (PhD diss., University of Oregon, 2013).

44. A report in from 1920 concluded fistulas were "more common following operations than following childbirth." Edward Starr Judd, "The Operative Treatment of Vesicovaginal Fistulae," *Surgery, Gynecology and Obstetrics* 30, no. 5 (May 1920): 447–453; Syed Z.H. Jafri, Ananias C. Diokno, and Marco A. Amendola, *Lower Genitourinary Radiology: Imaging and Intervention* (New York: Springer, 1997): 63.

45. van Deventer, *Manuale operatien.*

46. Chicago obstetrician Effa V. Davis argued that given the frequency of small or deformed pelvises, "the size of the infant can be regulated by diet and exercise if carried out strictly for a proper time." "A Study of the Bony Pelvis in One Hundred and Fifty Cases," *JAMA* 45, no. 23 (1905): 1709–10. Actually, obstetricians *could* enlarge the pelvis a bit by cutting the cartilage and ligaments between the pelvic bones. This operation, symphysiotomy, was painful and posed a difficult recovery, but had lower maternal mortality than cesareans in through the early twentieth century. It has a troubling history in Ireland, where thousands of women underwent the procedure, often without their knowledge. See Cara Delay, "'The Torture Began': Symphysiotomy and Obstetric Violence in Modern Ireland," *Nursing Clio*, May 31, 2016, https://nursingclio.org/2016/05/31/the-torture-began-symphysiotomy-and-obstetric-violence-in-modern-ireland/. Today the operation has advocates in areas of the world where access to hospitals capable of safe cesareans is especially challenging. A. Wilson, E. G. Truchanowicz, D. Elmoghazy, C. MacArthur, and A. Coomarasamy, "Symphysiotomy for Obstructed Labour: A Systematic Review and Meta-Analysis, *BJOG* 123 (2016):1453–61.

47. They are gynecoid, nearly circular, considered the normal female pelvic shape; android pelvises appear more like the typical male pelvis, slightly heart-shaped and narrow side-to-side; anthropoid describes an oval pelvis that is taller than it is wide; and the platypelloid or "flat." The system widely employed today, the Caldwell-Moloy classification, was developed in the 1930s. Today it faces increasing criticism, mostly for trying to lump the "nebulous cloudy continuum of shape variation" into four simple, or simplistic, categories. See Algis Kuliukas, Lesley Kuliukas, Daniel Franklin, and Ambika Flavel, "Female Pelvic Shape: Distinct Types or Nebulous Cloud?" *British Journal of Midwifery* 23, no. 7 (published online: July 1, 2015), https://doi.org/10.12968/bjom.2015.23.7.490.

48. In 1934, Herbert Thoms of the Yale School of Medicine questioned the very relationship between rickets and the "normal pelvis." He speculated that the "oval or female type pelvis [previously presumed to be "normal"] might be a result of mild rachitis occurring in early childhood or at puberty," and that improved diet and social conditions were reducing the

incidence of this type in favor of round pelvises, which predominate "in aboriginal people living in climates in which rachitis does not occur." Herbert Thoms, "Is the Oval or Female Type Pelvis a Rachitic Manifestation?" *American Journal of Obstetrics and Gynecology* 31, no. 1 (1936): 111–15.

49. J. Whitridge Williams, "The Frequency of Contracted Pelves in the First Thousand Women Delivered in the Obstetrical Department of the Johns Hopkins University," *Obstetrics* 1, nos. 5 and 6 (offprint, 1899). A Chicago obstetrician in 1905 found that 17 percent of his patients had "deformed" pelvises; he blamed diet and environment, especially "improper breast feeding," which he labelled "commonly the cause of rickets." Davis, "Bony Pelvis."

50. Edwin Bradford Cragin and George Hope Ryder, *Obstetrics: A Practical Text-Book for Students and Practitioners* (Philadelphia: Lea & Febiger, 1916), 648–65.

51. Cragin and Ryder, *Obstetrics*, 655. Williams, "Frequency of Contracted Pelves."

52. J. Whitridge Williams, "Pelvic Indications for the Performance of Cesarean Section," *American Medicine* 2, no. 13 (September 28, 1901): 483–88; all quotes from 484.

53. Williams provided some details about the one death, essentially blaming the patient, an African American, who "refused to enter the hospital." The operation took place "in a room so small that it was necessary to take down the bed in order to make room for the operating table." Williams, "Pelvic Indications," 486.

54. Williams, "Pelvic Indications," 488.

55. J. Whitridge Williams, "A Critical Analysis of Twenty-One Years Experience with Caesarean Section," *Transactions of the Medical Association of the State of Alabama, 1921* (Montgomery, AL: Brown Printing Company, 1921), 205–44.

56. Jacqueline Wolf, "Risk and Reputation: Obstetricians, Cesareans, and Consent," *Journal of the History of Medicine and Allied Sciences* 73, no. 1 (2018): 7–28, quote 8–9; CDC reported C-sections accounted for 31.7 percent of all live births in the US. Vital Statistics Rapid Release, no. 8 (May 2020), https://www.cdc.gov/nchs/data/vsrr/vsrr-8-508.pdf.

57. Jacqueline Wolf, *Cesarean Section: An American History of Risk, Technology, and Consequence* (Baltimore: Johns Hopkins University Press, 2018), 3.

58. Williams, "Critical Analysis," 235–36.

59. J. Whitridge Williams, *Obstetrics: A Text-Book for the Use of Students and Practitioners* (New York: D. Appleton, 1903); "any one with two hands . . . ," J. Whitridge Williams, "Cesarean Section at the Johns Hopkins

Hospital," *Northwest Medicine* 25 (October 1926): 519–26, quote 526. Quoted in Wolf, *Cesarean Section*, 5.

60. Caroline H. Bledsoe, "'Is Rickets a Normal Condition': The Construction of Normality in the Reproductive Body in Early and Late Twentieth Century America," in *Discovering Normality in Health and the Reproductive Body*, ed. Caroline Bledsoe PAS Working Papers, no. 11, Program of African Studies (Evanston, IL: Northwestern University, 2002): 51–62.

61. Researchers spent the first forty years of the x-ray era working out its technical kinks, praising its virtues, publishing dozens of studies, and expanding the sorts of pelvic measurements it could make, always touting its results as superior to instrument or manual measurements. Medicine's embrace of x-ray pelvimetry was hot but short. By the 1960s concerns with cancer risks, especially to fetuses exposed to x-rays, started eroding enthusiasm for the technology, and by the 1980s obstetricians rarely ordered x-ray pelvimetry studies, especially when new imaging technologies, from ultrasound to MRI and computed tomography, made their way into obstetrics; Edward Steward Taylor, *History of the American Gynecological Society, 1876–1981, and the American Associate of Obstetricians and Gynecologists, 1888–1981* (St. Louis: C.V. Mosby, 1985), 36; Herbert Thomas, "Roentgen Pelvimetry as a Routine Prenatal Procedure," *American Journal of Obstetrics and Gynecology* 40, no. 5 (1940): 891–905; see also Anja Hiddings and Stuart S. Blume, "Technology, Science, and Obstetric Practice: The Origins and Transformation of Cephalopelvimetry," *Science, Technology, & Human Values* 17, no. 2 (Spring 1992): 154–79, esp. 165–71.

62. The pelvimeter could provide general information about the size and shape of a pelvis but not, as a British obstetrician scolded, the detailed information needed "to decide on such grave issues as Caesarean section or even induction of labour." Walter S. A. Griffith, letter to the editor, re "Plea for Abolition of the Pelvimeter," *BMJ* 2, no. 3857 (December 1934): 1075.

63. Williams, "Critical Analysis"; Theodore R. George Jr. and John F. J. Clark, "The Changing Role of Cesarean Section," *Journal of the National Medical Association* 64, no. 3 (May 1972): 227–29. Freedman's Hospital's category "Repeat C/S" does not make clear why the initial cesarean was performed.

64. Williams, "Critical Analysis," 217.

65. The CDC put the rate for all cesarean sections at 31.7 percent in 2019; the rate for Black women was 35.9. CDC, *National Vital Statistics Reports* 70, no. 2 (March 23, 2021), https://www.cdc.gov/nchs/data/nvsr/nvsr70/nvsr70-02-508.pdf; "Overall pregnancy related mortality" was 17.2 deaths per 100,000 live births, but the rate for non-Hispanic Black

women was 43.5/100,000. Gianna Melilo, "Racial Disparities Persist in Maternal Morbidity, Mortality and Infant Health," AJMC.com, accessed November 3, 2023, https://www.ajmc.com/view/racial-disparities-persist -in-maternal-morbidity-mortality-and-infant-health.

66. Helen Barrett and Aidan McElduff, "Vitamin D and Pregnancy: An Old Problem Revisited," *Best Practice & Research Clinical Endocrinology & Metabolism* 24 (2010): 527–39, https://doi.org/10.1016/j.beem.2010.05.010.

Chapter 8

1. Greg Retsinas, "Couple Guilty of Assault in Vegan Case," *New York Times*, April 5, 2003, Section D, 1, https://www.nytimes.com/2003/04/05 /nyregion/couple-guilty-of-assault-in-vegan-case.html; "Doc Boosts Vegan Duo—Don't Blame Diet for Baby's Withered State, He Testifies," *New York Post*, March 26, 2003, Sport and Late City Final, 24; "Vitamin D Is Urged for Breast-Fed Babies," *New York Times*, April 7, 2003, Late Ed. Final, Section A, 14.

2. Herbert Lowe, "Baby-Starving Trial Begins: Defense: Parents Were 'Caring,'" *Newsday*, March 17, 2003, 9, http://www.newsday.com/news /baby-starving-trial-begins-defense-parents-were-caring-1.355744; Retsinas, "Couple Guilty"; Herbert Lowe, "Prison for Vegans; Judge Rips Parents of Infant Deprived of Proper Nutrition," *Newsday*, May 19, 2003, http://www.newsday.com/news/prison-for-vegans-judge-rips-parents-of -infant-deprived-of-proper-nutrition-1.256495.

3. Thomas R. Welch, William H. Bergstrom, and Reginald C. Tsang. "Vitamin D–Deficient Rickets: The Reemergence of a Once-Conquered Disease," *Journal of Pediatrics* 137, no. 2 (August 2000): 143–45, https:// doi.org/10.1067/mpd.2000.109008.

4. "Tragic Tot's 'Rotten' Diet: Vegan Parents Fed Her Spoiled Produce," *New York Post* May 2, 2002, 17; Scott Shifrel, "Was Tot's Vegan Diet Criminal?" *New York Daily News*, March 10, 2003, http://www.nydailynews.com/news /crime_file/story/65863p-61352c.html; Lowe, "Prison for Vegans."

5. For example, Retsinas, "Couple Guilty."

6. Russell W. Chesney, "Rickets: The Third Wave," *Clinical Pediatrics* 41 (April 2002): 137–39; Carvalho, Kenney, Carrington, and Hall, "Severe Nutritional Deficiencies," E46; Kay M. Tomashek, Shanna Nesby, Kelley Scanlon, Mary E. Cogswell, Kenneth Powell, Umesh Parashar, Anne Mellinger-Birdsong, Laurence Grummer-Strawn, William Dietz, "Nutritional Rickets in Georgia," *Pediatrics* 107, no. 4 (April 2001): e45.

7. Patrice Jones, Mark Lucock, Martin Veysey and Emma Beckett, "The Vitamin D–Folate Hypothesis as an Evolutionary Model for Skin

Pigmentation: An Update and Integration of Current Ideas," *Nutrients* 10, no. 5 (2018): 554, https://doi.org/10.3390/nu10050554.

8. Bodo Lehmann and Michael Meurer, "Vitamin D Metabolism," *Dermatologic Therapy* 23 (2010): 2–12.

9. Krista S. Crider, Lynn B. Bailey, and Robert J. Berry, "Folic Acid Food Fortification—Its History, Effect, Concerns, and Future Directions" *Nutrients* 3, no. 3 (2011): 370–384, https://doi.org/10.3390/nu3030370; Nina G. Jablonski, "The Evolution of Human Skin Pigmentation Involved the Interactions of Genetic, Environmental, and Cultural Variables," *Pigment Cell & Melanoma Research* 34, no. 4 (2021): 707–29.

10. Jablonski, "Evolution," 709.

11. Ewen Callaway, "Europe's First Humans: What Scientists Do and Don't Know," *Nature News Explorer* June 22, 2015, https://www.nature.com/news/europe-s-first-humans-what-scientists-do-and-don-t-know-1.17815#/2. *Homo sapiens* made at least one fleeting first start in Eurasia as early as 210,000 years ago, but within 50,000 years, Neanderthals replaced them. Permanent *homo sapiens* settlement in Eurasia came around 60,000 years ago.

12. Ze'ev Hochberg and Irit Hochberg, "Evolutionary Perspective in Rickets and Vitamin D," *Frontiers in Endocrinology* 10 (2019): 306, https://doi.org/10.3389/fendo.2019.00306.

13. Frederick G. Murray, "Pigmentation, Sunlight, and Nutritional Disease," *American Anthropologist*, New Series 36, no. 3 (1934), 438–45.

14. Lia Betti and Andrea Manica, "Human Variation in the Shape of the Birth Canal Is Significant and Geographically Structured," *Proceedings of the Royal Society B: Biological Sciences* 285 (2018): 20181807, http://dx.doi.org/10.1098/rspb.2018.1807.

15. Jablonski, "Evolution," 709.

16. George R. Milner, "Early Agriculture's Toll on Human Health," *Proceedings of the National Academy of Sciences* 116, no. 28 (July 2019): 13721–13723, https://doi.org/10.1073/pnas.1908960116; Pauline Trinh, Jesse R. Zaneveld, Sarah Safranek, and Peter M. Rabinowitz, "One Health Relationships between Human, Animal, and Environmental Microbiomes: A Mini-Review," *Frontiers in Public Health* 6 (August 30, 2018): 235, https://doi.org/10.3389/fpubh.2018.00235. When introduced into towns and cities, these new microbial threats regularly erupted in epidemics with mass mortality, providing in turn powerful selective advantages for individuals with partial immunity.

17. D. M. Paige, "Lactose Intolerance," in *Encyclopedia of Human Nutrition*, 2nd ed., Lindsay Allen and Andrew Prentice, eds. (Cambridge, MA: Academic Press, 2005), https://www.sciencedirect.com/referencework

/9780122266942/encyclopedia-of-human-nutrition; Benjamin Phelan, "The Most Spectacular Mutation in Recent Human History," *Slate*, October 23, 2012, https://slate.com/technology/2012/10/evolution-of-lactose-tolerance-why-do-humans-keep-drinking-milk.html.

18. One recent example of associating vitamin D and lactase persistence: "But if individuals are able to get calories and most nutritional benefits (including calcium and vitamin D) from dairy products without genetic adaptation." Laure Segurel, Perle Guarino-Vignon, Nina Marchi, Sophie Lafosse, Romain Laurent, Céline Bon, Alexandre Fabre, Tatyana Hegay, and Evelyne Heyer, "Why and When Was Lactase Persistence Selected For? Insights from Central Asian Herders and Ancient DNA," *PLOS Biology* 18, no. 6: e3000742, https://doi.org/10.1371/journal.pbio.3000742. Though milk did not provide much vitamin D, it did provide calcium, and lactose can help absorb calcium, giving vitamin D a small boost. Andrea S Wiley, "Lactose Intolerance," *Evolution, Medicine, and Public Health* 2020, no.1 (2020): 47–48, https://doi.org/10.1093/emph/eoaa006.

19. Nina Jablonski, "Skin Color and Race," *American Journal of Physical Anthropology* 175 (2021): 437–47, https://onlinelibrary.wiley.com/doi/full/10.1002/ajpa.24200; Emmanuel Chukwudi Eze, "The Color of Reason: The Idea of 'Race' in Kant's Anthropology," in Eze, ed., *Postcolonial African Philosophy: A Critical Reader* (Cambridge, MA, Blackwell, 1997), 118; Darwin was hardly antiracist, but he was adamant about all humans being one species with a common origin. In *Descent of Man* he provided a sharp but understated takedown of those determined to classify the races: "But the most weighty of all the arguments against treating the races of man as distinct species, is that they graduate into each other . . . Man has been studied more carefully than any other animal, and yet there is the greatest possible diversity amongst capable judges whether he should be classed as a single species or race, or as two (Virey), as three (Jacquinot), as four (Kant), five (Blumenbach), six (Buffon), seven (Hunter), eight (Agassiz), eleven (Pickering), fifteen (Bory de St-Vincent), sixteen (Desmoulins), twenty-two (Morton), sixty (Crawfurd), or as sixty-three, according to Burke. Jim Bindon, "Darwin's Borrowed Allegory and the Apocryphal Six Races of Buffon," accessed July 29, 2021, http://jbindon.people.ua.edu/race-and-human-variation.

20. Ellsworth Huntington, *Civilization and Climate* (New Haven, CT: Yale University Press, 1915); historian Mia Bay notes in a survey of Afrocentrism, that Jeffries is "widely quoted" but "not widely cited." Mia Bay, "The Historical Origins of Afrocentrism," *Amerikastudien / American Studies* 45, no. 4 (2000): 501–12, 505n12.

21. Acker, "Rickets in Negroes"; "Have We Any Rickets?"

22. Downs, *Sick from Freedom*.

23. Williams, "Critical," 217. Wallace Craig and Morris Belkin, "The Prevention and Cure of Rickets," *Scientific Monthly* 20, no. 5 (May 1925): 541–50, 542.

24. Alfred Hess and S. N. Blackberg, "An Experimental Study of 'The Constitutional Factor' in the Etiology of Rickets," *American Journal of Physiology* 102, No. 1 (September 30, 1932): 8.

25. "dark skinned races," Wilson, "Food and Health"; "physical deterioration," Snow, "Explanation," 161.

26. P. D. Saville and M. H. Alderman, "Deficiency Rickets in New York: Dissociation Between Urinary Hydroxyproline and Glycylproline with Treatment," *Archives of Internal Medicine* 125, no. 2 (1970): 341–43, quote 342; B. N. Ames, W. B. Grant, W. C. Willett, "Does the High Prevalence of Vitamin D Deficiency in African Americans Contribute to Health Disparities?" *Nutrients* 13 (2021): 499, https://doi.org/10.3390/ nu13020499.

27. Bivins, "'English Disease'"; Paul Lips and Renate T. de Jongh, "Vitamin D Deficiency in Immigrants," *Bone Reports* 9 (June 11, 2018): 37–41, https://doi.org/10.1016/j.bonr.2018.06.001; E. Mallet, J. Gaudelus, P. Reinert, B. Luyer, C. Lecointre, J. Léger, C. Loirat, B. Quinet, J. Bénichou, J. Furioli, G. A. Loeuille, B. Roussel, M. Larchet, F. Freycon, M. Vidailhet, and I. Varet, "Symptomatic Rickets in Adolescents," *Archives de Pédiatrie: Organe Officiel de la Sociéte Française de Pédiatrie* 11 (2004): 871–8, 10.1016/ S0929693X04002829; Tom D. Thacher, Pawel Pludowski, Nick J. Shaw, M. Zulf Mughal, Craig F. Munns, and Wolfgang Högler, "Nutritional Rickets in Immigrant and Refugee Children," *Public Health Reviews* 37, no. 3 (July 22, 2016), https://doi.org/10.1186/s40985-016-0018-3.

28. Lips and de Jongh, "Vitamin D Deficiency."

29. A. K. Bhattacharyya, "Nutritional Rickets in the Tropics," *World Review of Nutrition and Dietetics* 67 (1992): 140–97, cited in Tom Thacher, Philip R. Fischer, Mark Strand, and John M. Pettifor, "Nutritional Rickets around the World: Causes and Future Directions," *Annals of Tropical Paediatrics* 26 (2006): 1–16.

30. Thacher, Fischer, Strand, Pettifor, "Nutritional Rickets around"; Tom Thacher, Philip R. Fischer, C. O. Isichei, A. I. Zoakah, and John M. Pettifor, "Prevention of Nutritional Rickets in Nigerian Children with Dietary Calcium Supplementation," *Bone* 50, no. 5 (May 2012): 1074–80, https://doi.org/10.1016/j.bone.2012.02.010, PMID: 22373953.

31. Frank R. Greer, "Vitamin D Deficiency—It's More than Rickets," *Journal of Pediatrics* 143, no. 4, (October 1, 2003): 422–23, https://doi .org/10.1067/S0022-3476(03)00465-7.

32. This is an example, perhaps of a transitional terminology; see
J. M. Pettifor, J. M. Isdale, J. Sahakian, and J. D. Hansen, "Diagnosis of
Subclinical Rickets," *Archives of Disease in Childhood* 55, no. 2 (1980):
155–57, https://doi.org/10.1136/adc.55.2.155.

33. Edwards A. Park, "The Use of Vitamin D Preparations in the Prevention and Treatment of Disease," *JAMA* 111, no. 13 (1938):1179–87,
https://doi.org/10.1001/jama.1938.72790390002011.

34. The goal was not to measure vitamin D itself, but one of its building blocks, 25-hydroxy vitamin D. The first assays, developed in the early
1970s, were cumbersome and expensive, useful only for research laboratories. Bruce W. Hollis, "Editorial: The Determination of Circulating
25-Hydroxyvitamin D: No Easy Task," *Journal of Clinical Endocrinology &
Metabolism* 89, no. 7 (July 1, 2004): 3149–51, https://doi.org/10.1210/jc
.2004-0682.

35. Michael F. Holick, "Vitamin D Status: Measurement, Interpretation
and Clinical Application," *Ann Epidemiol* 19, no. 2 (February 2009): 73–78,
https://doi.org/10.1016/j.annepidem.2007.12.001.

36. NHANES III (1988–1994) was the first of the national sur-
veys to perform vitamin D assays in the mix. Ashraf Zadshir, Naureen
Tareen, Deyu Pan, Keith Norris, and David Martins, "The Prevalence of
Hypovitaminosis D among US Adults: Data from the NHANES III," *Ethnicity
& Disease* 15 (Autumn 2005): S5-97-101.

37. On the history of vitamin D's role in treating osteoporosis, see
Gerald Grob, *Aging Bones: A Short History of Osteoporosis* (Baltimore: Johns
Hopkins University Press, 2014).

38. William B. Grant, Heide S. Cross, Cedric F. Garland, Edward D.
Gorham, Johan Moan, Meinrad Peterlik, Alina C. Porojnicu, Jörg Reichrath,
and Armin Zittermann, "Estimated Benefit of Increased Vitamin D Status
in Reducing the Economic Burden of Disease in Western Europe," *Progress
in Biophysics and Molecular Biology* 99, no. 2–3 (2009): 104–13, ISSN 0079-
6107, https://doi.org/10.1016/j.pbiomolbio.2009.02.003.

39. Pierre Olivier Lang and Richard Apinall, "Vitamin D Status and
the Host Resistance to Infections: What Is Currently (Not) Understood,"
Clinical Therapeutics 39, no. 5 (May 1, 2017): 930–45, https://www.ncbi.nlm
.nih.gov/pmc/articles/PMC6695859/.

40. Lang and Apinall, "Vitamin D Status," 932.

41. Simiao Chen, Klaus Prettner, Michael Kuhn, Pascal Geldsetzer, Chen
Wang, Till Bärnighausen, and David E. Bloom, "Climate and the Spread
of COVID-19," *Scientific Reports* 11, article number 9042 (April 27, 2021),
https://www.nature.com/articles/s41598-021-87692-z.

42. Suzanne Elvidge, "Can Vitamin D Help Protect against COVID?" *Scientific American*, June 24, 2021, https://www.scientificamerican.com /article/can-vitamin-d-help-protect-against-covid/; Philip Bump, "Yet Again, Trump Pledges that the Coronavirus Will Simply Go Away," *Washington Post*, April 28, 2020, https://www.washingtonpost.com/politics/2020/04/28 /yet-again-trump-pledges-that-coronavirus-will-simply-go-away/.

43. "Study Suggests High Vitamin D Levels May Protect against COVID-19, Especially for Black People," March 29, 2021, University of Chicago Department of Medicine, https://medicine.uchicago.edu/study -suggests-high-vitamin-d-levels-may-protect-against-covid-19-especially -for-black-people/; "Vitamin D May Not Protect against COVID-19, as Previously Suggested," *McGill Newsroom*, June 4, 2021, https://www.mcgill .ca/newsroom/channels/news/vitamin-d-may-not-protect-against-covid-19 -previously-suggested-331358.

44. J. E. Manson, S. S. Bassuk, "Commentary. Eliminating Vitamin D Deficiency During the COVID-19 Pandemic: A Call to Action," *Metabolism: Clinical and Experimental* 112, 154322, advance online publication, https:// doi.org/10.1016/j.metabol.2020.154322.

45. I have been deliberately imprecise in my terminology, folding specific measure and specific vitamin D metabolites used in determin- ing vitamin D status. In this instance, "20 ng/ml" refers to 20 nanograms 25-hydroxyvitamin D (25OHD) per milliliter blood serum; (25(OH)D) is the primary circulating form of vitamin D in the blood.

46. The IOM is the branch of the National Academies now known as the National Academy of Medicine. Institute of Medicine, *Dietary Reference Intakes for Calcium and Vitamin D* (Washington, DC: National Academies Press, 2011).

47. Robert P. Heaney and Michael F. Holick, "Why the IOM Recommendations for Vitamin D are Deficient," *Journal of Bone and Mineral Research: The Official Journal of the American Society for Bone and Mineral Research* 26, no. 3 (2011): 455–57, https://doi.org/10.1002/jbmr.328.

48. Michael P. Horan, Kevin Williams, and Daniel Hughes, "The Role of Vitamin D in Pediatric Orthopedics," *Orthopedic Clinics of North America* 50 (2019): 181–91, https://doi.org/10.1016/j.ocl.2018.10.002.

49. JoAnne E. Manson, et al., "Vitamin D Supplements and Prevention of Cancer and Cardiovascular Disease, *New England Journal of Medicine* 380, no. 1 (January 3, 2019): 33–44, quote 39.

50. Victor Herbert, "The Vitamin Craze," *Archives of Internal Medicine* 140, no. 2 (1980): 173–76, https://doi.org/10.1001/archinte.1980.0033014 0031014. Despite his well-earned reputation as a harsh critic of nutrition quackery, Herbert was a Columbia University–trained hematologist whose

early research focused on B9 (folate) and B12. He famously demonstrated how folate deficiency developed by putting himself on a diet with almost no folate. He nearly died in the experiment, but his findings produced a landmark report. John M. Scott, "Victor Herbert (1927–2002)," *Journal of Nutrition* 134, no. 7 (July 2004): 1678–80, https://doi.org/10.1093/jn/134.7.1678.

51. JoAnne Manson and SS Bassuk, "Vitamin and Mineral Supplements: What Clinicians Need to Know," *JAMA* 319, no. 9 (2018): 859–60, https://doi.org/10.1001/jama.2017.21012.

52. Reports and Data, "Clinical Laboratory Services Market Analysis, By Specialty," accessed August 21, 2021, https://www.reportsanddata.com/report-detail/clinical-laboratory-services-market; Transparency Marketing, "Vitamin D Testing Market," accessed August 21, 2021, https://www.transparencymarketresearch.com/vitamin-d-testing-market.html; Liz Szabo, "Vitamin D, the Sunshine Supplement, Has Shadowy Money Behind It," *New York Times*, August 18, 2018, https://www.nytimes.com/2018/08/18/business/vitamin-d-michael-holick.html.

53. Vera Tweed, "Vitamin D: How Much Is Enough?" *Better Nutrition* 82, no. 2 (February 2020): 14.

54. Szabo, "Sunshine Supplement."

55. DeLuca, "History."

56. His notoriety extends beyond potential economic entanglements. In 2018, *The New Yorker* ran an exposé on Holick's role as expert witness on behalf of accused parents in child abuse cases. David Armstrong, "The Child-Abuse Contrarian," *The New Yorker*, September 26, 2018.

57. Holick's website, accessed November 3, 2023, is http://drholick.com/.

58. Szabo, "Sunshine Supplement."

59. For example from the midcentury, see Apple, *Vitamania*, 127–29.

60. For example see Christie Aschwanden, "How Much Vitamin D Do You Need to Stay Healthy?" *Scientific American*, January 1, 2024, https://www.scientificamerican.com/article/how-much-vitamin-d-do-you-need-to-stay-healthy/. Aschwanden starts off declaring the end of the vitamin D era: After its rise from the 1930s vitamin D came to be seen as a miracle drug. "Then the bottom fell out. Although thousands of studies had linked low levels of vitamin D to an assortment of medical conditions, when scientists tried administering it as a means to prevent or treat those problems, the wonder supplement failed miserably." Of course, she gives voice to both the naysayers and the cheerleaders, like Michael Holick, whom she paints as firmly in the pocket of the vitamin industry. If you read around the sarcasm, Aschwanden makes the case put succinctly in her closing sentence: "The relation between the vitamin and disease is far more

complicated and nuanced than it first seemed and a reminder that scientific understanding is always evolving."

61. Joseph Garland, "Rickets" *Journal of the National Medical Association* 17, no. 3 (1925): 125–28.

62. Murray, "Pigmentation, Sunlight."

63. Peter Frost, "To Supplement or Not to Supplement: Are Inuit Getting Enough Vitamin D?" *Études Inuit Studies* 40, no. 2 (2016): 271–91, https://doi.org/10.7202/1055442ar; Peter Frost, "Vitamin D Deficiency among Northern Native Peoples: A Real or Apparent Problem?" *International Journal of Circumpolar Health* 71 18001 (March 19, 2012), https://doi.org/10.3402/IJCH.v71i0.18001; but see the response from a scientist whose conflict of interest statement reads "I receive funding from the UV Foundation . . . , Bio-Tech Pharmacal . . . , the Vitamin D Council . . . and the Vitamin D Society." William B. Grant, "Re: Vitamin D Deficiency among Northern Native Peoples," *International Journal of Circumpolar Health*, 71, 18434, https://doi.org/10.3402/ijch.v71i0.18434.

64. LaVerne L. Brown, Barbara Cohen, Derrick Tabor, Giovanna Zappalà, Padma Maruvada, and Paul M. Coates, "The Vitamin D Paradox in Black Americans: A Systems-Based Approach to Investigating Clinical Practice, Research, and Public Health—Expert Panel Meeting Report," *BMC Proceedings* 12, supplement 6 (May 9, 2018): 6, https://doi.org/10.1186 /s12919-018-0102-4.

65. Camille E. Powe, Michele K. Evans, Julia Wenger, Alan B. Zonderman, Anders H. Berg, Michael Nalls, Hector Tamez, Dongsheng Zhang, Ishir Bhan, S. Ananth Karumanchi, Neil R. Powe, and Ravi Thadhani, "Vitamin D-Binding Protein and Vitamin D Status of Black Americans and White Americans," *New England Journal of Medicine* 369, no. 21 (November 21, 2013): 1991–2000, https://doi.org/10.1056/NEJM oa1306357, PMID: 24256378; PMCID: PMC4030388; quote Amy Norton, "Vitamin D Deficiency Might Be Overdiagnosed in Blacks, Study Suggests," *Newsday*, November 20 2013, https://www.newsday.com/news/health /vitamin-d-deficiency-might-be-overdiagnosed-in-blacks-study-suggests -e72249.

66. Carvalho, Kenney, Carrington, and Hall, "Severe Nutritional Deficiencies," E46.

67. Michael Yudell, Dorothy Roberts, Rob Desalle, and Sarah Tishkoff, "Taking Race Out of Human Genetics," *Science* 351, no. 6273 (February 5, 2016): 564–65.

68. Holick, "Vitamin D Status."

69. Marietta Eichelberger, director, Nutrition Services, Irradiated Evaporated Milk Institute (Chicago) to Dr. J. C. Gittings, Un Hospital Phila,

October 25, 1934, Folder "Milk, Wisconsin Fund" F #2, Joseph Stokes Papers, American Philosophical Society B: St65P/.

70. Hess quote, "Milk Irradiation for Rickets," 17.

71. Kevin N. Laland, John Odling-Smee, and Sean Myles, "How Culture Shaped the Human Genome: Bringing Genetics and the Human Sciences Together," *Nature Reviews Genetics* 11 (2010): 137–148, https://doi.org/10 .1038/nrg2734.

72. Theresa A. Nicklas, Haiyan Qu, Sheryl O. Hughes, Sara E. Wagner, H. Russell Foushee, and Richard M. Shewchuk, "Prevalence of Self-reported Lactose Intolerance in a Multiethnic Sample of Adults," *Nutrition Today* 44, no. 5 (September 2009): 222–27, https://doi.org/10.1097/NT.0b013e 3181b9caa6.

73. Typical tests administered fifty grams of lactose in water, four times the lactose found in an eight-ounce glass of milk. Claire L. Jansson-Knodell, Edward J. Krajicek, Dennis A. Savaiano, and Andrea S. Shin, "Lactose Intolerance: A Concise Review to Skim the Surface," *Mayo Clinic Proceedings* 95, no. 7 (2020): 1499–1505, https://doi.org/10.1016/j.mayocp.2020.04.036.

74. Jeanette N. Keith, "Lactose Intolerance and Milk Protein Allergy," *Current Treatment Options in Gastroenterology* 18 (2020): 1–14.

Chapter 9

1. Chris Orchard, "Genomic Medicine in the Real World: 'Hope' and 'Hype,'" Harvard T. H. Chan School of Public Health," accessed August 1, 2022, https://www.hsph.harvard.edu/ecpe/genomic-medicine-in-the-real -world-hope-and-hype/.

2. Julia Belluz, "I Asked 8 Researchers Why the Science of Nutrition Is So Messy. Here's What They Said," *Vox*, August 16, 2016, https://www.vox .com/2016/1/14/10760622/nutrition-science-complicated; Edward Archer, Carl J. Lavie, and James O. Hill, "The Failure to Measure Dietary Intake Engendered a Fictional Discourse on Diet-Disease Relations," *Frontiers in Nutrition* 5, no. 105 (November 13, 2018), https://doi.org/10.3389/fnut .2018.00105.

3. Magdalini Patseadou and Dagmar M. Haller, "Vitamin D in Adolescents: A Systematic Review and Narrative Synthesis of Available Recommendations," *Journal of Adolescent Health* 66, no. 4 (April 2020): 388–407, https://doi.org/10.1016/j.jadohealth.2019.08.025.

4. Robert Ewart Ramsay, quoted in Dick, *Rickets*, 63. Ramsay was reporting on the records of the Los Angeles County Hospital for the previous seven years, which listed only twenty-eight cases of rickets diagnoses in a hospital with average annual admissions of fifteen thousand.

5. Skin Cancer Foundation, "Vitamin D & Sun Protection: What You Need to Know," accessed October 18, 2023, https://www.skincancer.org /skin-cancer-prevention/sun-protection/vitamin-d; Anahad O/Connor, "Low Vitamin D Levels Linked to Disease in Two Big Studies," *Well, New York Times* blog, April 1, 2014, https://well.blogs.nytimes.com/2014/04/01 /low-vitamin-d-levels-linked-to-disease-in-two-big-studies/.

6. Howard Frumkin et al., "Nature Contact and Human Health: A Research Agenda," *Environmental Health Perspectives* 125, no. 7 (July 31, 2017): 075001, https://doi.org/10.1289/EHP1663.

7. Rob Nixon, *Slow Violence and the Environmentalism of the Poor* (Cambridge, MA: Harvard University Press, 2011), 13.

8. In a recent interview, Nixon updates his take on the disproportionate impact on women, Aram Goudsouzian, "Slow Violence, Then and Now," *Chapter 16*, November 10, 2021, https://chapter16.org/slow-violence-then -and-now/; quote ("not just attritional") Nixon, *Slow Violence*, 3.

9. Jill S. Schneiderman, "Awake in the Anthropocene," *Contemporary Buddhism* 13, no. 1 (May 1, 2012): 83–97, quotes 89–90, https://doi.org /10.1080/14639947.2012.669283.

10. Saville and Alderman, "Deficiency Rickets," 342; Ames, Grant, and Willett, "High Prevalence," 499; or as Buffalo physician Irving Snow put it over a century earlier "a sign of "the physical deterioration of a southern race in a northern climate," Snow, "Explanation," 161.

11. René Dubos, *So Human an Animal: How We Are Shaped by Surroundings and Events*, with a new introduction by Jill Cooper & David Mechanic (New Brunswick: Transaction, 1998), 16.

Index

Page numbers followed by "f" refer to figures.

Medici family, 22–23

Medline, 195

melanin, 182, 183, 187, 199

Mellanby, Edward, 82–83, 84, 85, 244n41; career, 82; demonstrates rickets is deficiency disease, 83; on diet vs. environment as rickets cause, 85

Mellanby, May (née Tweedy), 82

Memphis, Tennessee, 64

Mendel, Lafayette, 81

Mercy Hospital (Buffalo, NY), 146

metabolized milk. *See under* vitamin D milk

miasma theory of health, 35, 75, 86

"microbe hunters," 77

Middlesex Hospital, 78

midwifery, physicians' encroachment on, 6, 162

migration: and clothing, 184; forced, 208; post–World War II, 188; prehistoric, 180–84; since the Bronze Age, 184–85; and technology, 184

migrations, from Africa, 182–83

milk: canned, 97; cow's, vitamin D content, 2; evaporated, 98–99, 122; evaporated, advertising, 99; evaporated, in rickets studies, human, 97; evaporated, vitamin D fortified, 68; falling consumption, 15; "Got Milk?" ad campaign, 15; human, fortification by irradiation, 131, 133f; "natural" vehicle for vitamin D fortification, 129, 202; as "Nature's Perfect Food," 129; presumed antirachitic, 9; production, 1; reputation as unhealthy, 129; in rickets studies, animal, 82, 87; in

rickets studies, human, 90, 92, 98. *See also* dairy, falling consumption; dairy industry; vitamin D milk

Mills, C. W., 71

Milton (Blake), 29

Mitchell, Thomas, 64

Modern Brewer, 123, 258n64

M'Ruer, Daniel, 112

Mumbai, 87

mummies, 21, 222nn8–10

Murray, Frederick, 198

"Muscular Christianity," 52

Nash, Ogden, 52

Nation, The, 93, 100

National Academy of Science, 154

"National cod liver oil compound," 156

National Institutes of Health, 145

National League, 91

Nature, 85

Negroes Burial Ground, 43–44

Newark, New Jersey, 147–48

New Deal, 160

New England Journal of Medicine, 101, 193

Newfoundland, 111; fisheries dispute of 1890, 252n5

New Haven, Connecticut, 94

New Orleans, 100

New Orleans Times, 114

New Public Health, 13, 130

Newsday, 178

New South, 63

New Theory for Obstetricians and Midwives, The (Deventer), 165

New York Administration of Children's Services, 178

New York City Board of Health, 134–35